ELECTRONICS DRAFTING WORKBOOK

FOURTH EDITION

Cyrus Kirshner, Ph.D.

Professor of Engineering
Los Angeles Valley College
Van Nuys, California

Kurt M. Stone

Senior Engineering Designer
ITT Gilfillan
San Fernando, California

GLENCOE

Macmillan/McGraw-Hill

Lake Forest, Illinois
Columbus, Ohio
Mission Hills, California
Peoria, Illinois

Equipment Used in Electronics Drafting

The equipment used in electronics drafting is, for the most part, the same as that used in mechanical drafting, such as:

- Drawing board
- 24" T square
- 45°/90° triangle
- 30°/60° triangle
- Compass
- X-acto knife (for printed circuit tapes)
- Protractor
- Scale (12" preferred with both 1/32" and 1/50" increments)
- French curve
- Erasing shield
- Dusting brush
- Drafting (or masking) tape
- Ruby drafting eraser
- Drafting pencils (H, 2H, and 4H) automatic preferred
- Sandpaper pad or file

The above list is the usual equipment required for mechanical drafting, which is the same as required for electronics drafting. The student having these items but in different sizes need not purchase any new equipment other than electronics templates:

Electronics template ANSI Y32.2
Logic template MIL-STD-806 (Y32.14) three-quarter size

Sponsoring Editor: Paul Berk
Editing Supervisor: Alfred Bernardi
Design and Art Supervisor: Nancy Axelrod
Production Supervisor: Al Rihner

Cover Designer: Nancy Axelrod
Cover Illustration: Roger Roth

Library of Congress Cataloging in Publication Data

Kirshner, Cyrus.
Electronics drafting workbook.

1. Electronics drafting. I. Stone, Kurt M.
II. Title.
TK7866.K57 1985 621.381'0221 84-26104
ISBN 0-07-034907-X

Electronics Drafting Workbook, Fourth Edition

Imprint 1992

Send all inquiries to: Glencoe Division, Macmillan/McGraw-Hill, 936 Eastwind Drive, Westerville, Ohio 43081.

5 6 7 8 9 10 11 12 13 14 15 EDW 00 99 98 97 96 95 94 93 92

ISBN 0-07-034907-X

CONTENTS

APPENDIXES

We respectfully dedicate this book to our families and our friends, and to the advancement of technical education.

PREFACE

The fourth edition of *Electronics Drafting Workbook* was written in the same format and style as the very successful first three editions. The authors' main purpose in this edition is to update this workbook with the latest practices being adopted by industry--primarily, the trend toward CAD (computer-aided drafting/design). This fourth edition represents a 36 percent increase in workbook pages. The new pages deal, for the most part, with CAD.

The new lessons being introduced are (1) hybrid microcircuits and (2) CAD. Hybrid microcircuit fabrication has become an important packaging technique in the continuing trend toward microminiaturization. CAD, as part of the CAD/CAM (computer-aided manufacturing) trend, is less tedious than manual drafting and a far more efficient means of filing and producing drawings.

The authors' approach to CAD instruction is unique because:

1. Students do not need a computer graphics system to learn the essence of CAD when using this workbook.
2. Many students (with sufficient background) and most teachers will find that CAD can be taught from this workbook without an accompanying text.
3. The CAD chapter contains many student exercises, and there is a matching appendix for permanent reference.
4. The simulated CAD routines cover most electronics drafting requirements, plus manual drafting and elements of CAD equipment design.
5. The CAD equipment design section includes, in a few pages, some explanation of the CAM aspects of design, the use of portable calculators, algebra, and trigonometry--all demonstrated in very easy-to-follow text, with similar exercises. These design considerations are typical parts of the solution of a CAD/CAM problem. Many students are pleasantly surprised to see the applicability of the mathematics they have studied.
6. Finally, a typical term project in CAD is demonstrated. Instructors can modify this project as they see fit.

There are about 225 lab hours of material to the fourth edition, which at Los Angeles Valley College is more than two semesters of work. Consequently, most teachers with similar time restrictions will have to pick and choose lessons to be omitted. To aid teachers in this decision, a table of drawing times, indexed by problem/page numbers, has been included in the *Solutions Manual*.

The book is written for students in high school and college, usually the last two years of high school or the first two years of college. It can also be used easily in adult education programs and by correspondence or home-study students. This workbook is largely based on the electronics drafting course that has been taught at both

Los Angeles Valley College since 1959 (for both day and evening classes), and at the West Valley Occupational Center, Woodland Hills, California. It has met with a great deal of success in both places. The course at Los Angeles Valley College is credit-transferable to California Polytechnic Institute (which is fully accredited) and meets the electronics drafting requirements for the electronics engineer's degree.

The sequence of lessons was designed with these considerations in mind:

1. To progress from a simpler, or more familiar, kind of drawing to one that is slightly more complex. Therefore, where practical, it would be well to follow the exercises in order.
2. To start the student out immediately with electronics drafting, without spending too much time on conventional mechanical drafting review, and to develop a consecutive interest in electronics drafting concepts.
3. To present the most frequently used types of drawings first and infrequently used types of drawings last (i.e., based on the experiences of most beginning drafters).
4. To bring together in a term project the many acquired drafting skills, or at least to indicate how this can be done.

The basic strategy of instruction used here is to present a little theory and an exercise on each page, with a step-by-step, "learn by doing" approach. Another strategy is to develop the habit of looking up new information (in the appendix, catalogs, etc.) and reviewing old information (from previous exercises, drawings, etc.).

The authors wish to thank the many instructors, engineers, drafting supervisors, and companies who contributed to the preparation of this book. In particular, we would like to mention:

Jack Bernard, Colorado Technical College, Colorado Springs, Colorado

James J. Bots, Meramec Community College, St. Louis, Missouri

Joseph Cutrone, Island Drafting and Technical Institute, Amityville, New York

ITT Gilfillan, San Fernando, California

The General Electric Company, for the FM tuner schematic diagram (page 27) and the power supply schematic diagram (page 55)

Edgar Hund, Los Angeles Pierce College, Woodland Hills, California

Wayne Michie, Virginia Western Community College, Roanoke, Virginia

William Moore, Tidewater Community College, Portsmouth, Virginia

James Poulson, Shasta College, Redding, California

RCA, for the color and black and white TV receiver block diagrams (page 20)

Jerry Rye, Evergreen Valley Community College, San Jose, California

The Zenith Electric Company for the color TV block diagram (page 22)

We would also like to express sincere appreciation to the following firms which contributed to, and are represented in, our appendixes:

Alpha Wire Corp., Arco Electronics, Inc., Department of Defense Military Standards, General Electric Company, Grayhill, Inc., Herman H. Smith, Inc., Hughes Semiconductors, Littlefuse Incorporated, Ohmite Mfg. Co., Texas Instruments Incorporated, Triad Transformer Corporation, USECO, Vitramon, Inc., and Westinghouse Electric Corporation.

For the fourth edition material we wish to thank and acknowledge the following:

Andromeda Systems Inc., Canoga Park, California
Engineering Systems Corp., Baton Rouge, Louisiana
Pacesetter Systems Inc., Sylmar, California

Cyrus Kirshner
Kurt M. Stone

Electromechanical lettering need not be artistic, but it should be neat and legible.

NOTES:

1) ELECTRONICS LETTERING IS LARGELY DONE FREEHAND ON VELLUM WITH OR AGAINST A 1/8 OR 1/10 GRID BACKGROUND.

2) THE LETTERING CAN BE EITHER VERTICAL OR *INCLINED*.

3) THE STUDENT SHOULD LEARN BOTH STYLES. HOWEVER, IT IS EASIER TO LEARN VERTICAL LETTERING FIRST AND *INCLINED LETTERING 2ND*, RATHER THAN VICE VERSA.

4) PRACTICE DEVELOPS TECHNIQUE.

Copy the **notes** taken from column 1 started below, going from bottom to top.

4) PRACTICE

1) ELECTRONICS LETTERING IS LARGELY

NOTES:

Copy in column 4 the entries of column 3 using 1/8" high vertical lettering.

$2\frac{1}{2}$ $3\frac{3}{4}$ $4\frac{5}{64}$ $5\frac{7}{64}$

$6\frac{7}{8}$ $7\frac{9}{32}$ $8\frac{9}{16}$ $9\frac{1}{10}$

$5.110^{\pm.005}$ DIA. $6.120^{+.002}_{-.000}$

TOL. = TOLERANCE

FRAC. = ±1/64; X° = ±1/2°

.XX = ±.03; .XXX = ±.010

2.250
2.251

CLEARANCE FIT

2.248
2.247

2.2519
2.2513

INTERFERENCE FIT

2.2500
2.2506

TITLE MECHANICAL DRAFTING REVIEW - LETTERING

DWG. NO. MDR-1

NAME	DATE	COURSE	GRADE	SCALE	SHEET 1 OF 2	PAGE 1

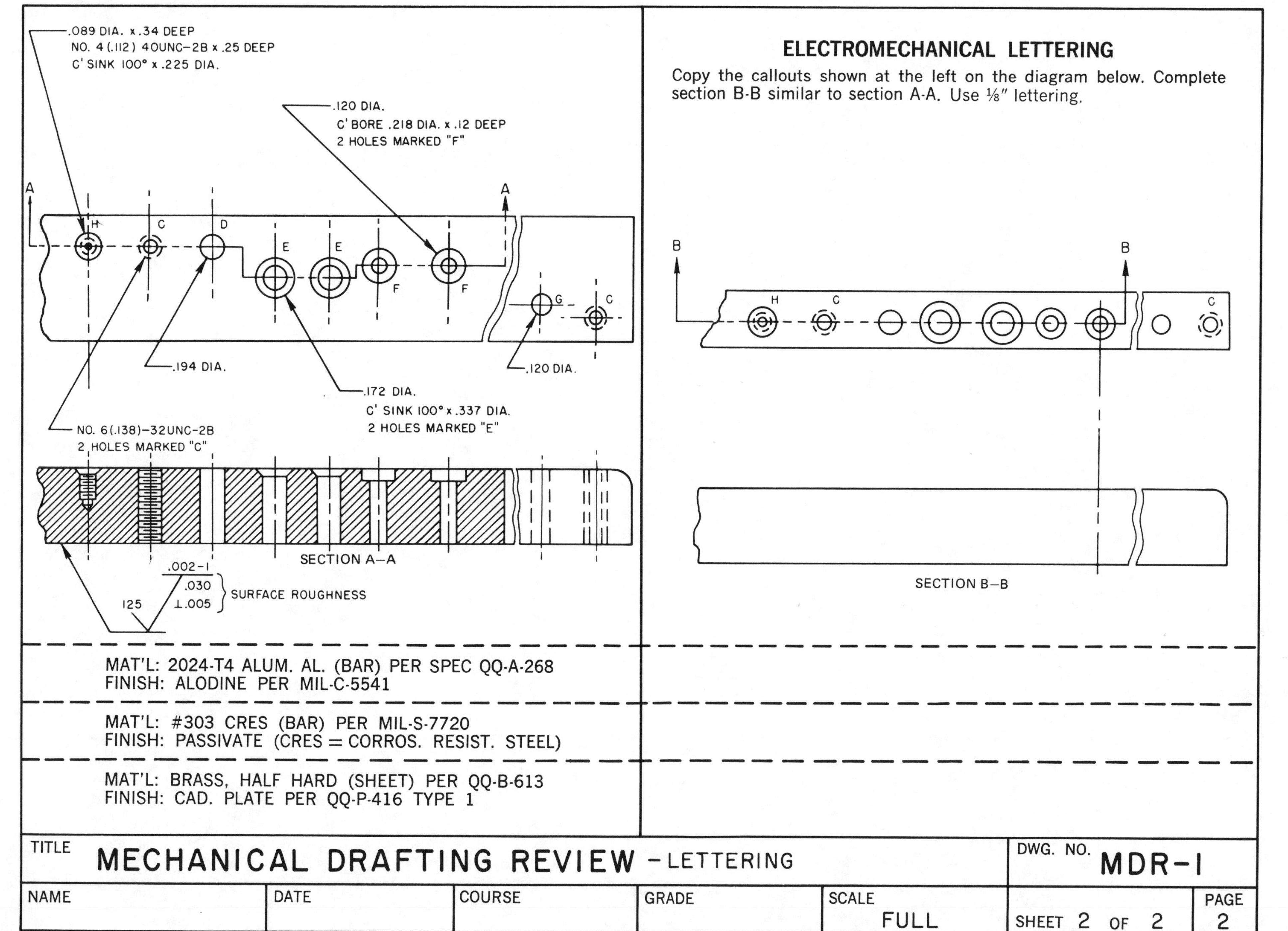
.089 DIA. x .34 DEEP
NO. 4 (.112) 40UNC–2B x .25 DEEP
C' SINK 100° x .225 DIA.
.120 DIA.
C' BORE .218 DIA. x .12 DEEP
2 HOLES MARKED "F"
A
A
H
C
D
E
E
F
F
G
C
.194 DIA.
.120 DIA.
.172 DIA.
C' SINK 100° x .337 DIA.
2 HOLES MARKED "E"
NO. 6 (.138)–32UNC–2B
2 HOLES MARKED "C"
SECTION A–A
.002–1
.030
⊥.005
125
SURFACE ROUGHNESS
ELECTROMECHANICAL LETTERING
Copy the callouts shown at the left on the diagram below. Complete section B-B similar to section A-A. Use 1/8" lettering.
B
B
H
C
C
SECTION B–B
MAT'L: 2024-T4 ALUM. AL. (BAR) PER SPEC QQ-A-268
FINISH: ALODINE PER MIL-C-5541
MAT'L: #303 CRES (BAR) PER MIL-S-7720
FINISH: PASSIVATE (CRES = CORROS. RESIST. STEEL)
MAT'L: BRASS, HALF HARD (SHEET) PER QQ-B-613
FINISH: CAD. PLATE PER QQ-P-416 TYPE 1
TITLE
MECHANICAL DRAFTING REVIEW – LETTERING
DWG. NO. MDR-1
NAME
DATE
COURSE
GRADE
SCALE FULL
SHEET 2 OF 2
PAGE 2

DIMENSIONING AND LINE WEIGHTS

VISIBLE LINE: thick (H lead)
For outline and cutting planes.

HIDDEN LINE: medium (2H lead)

SECTION LINE: thin (4H lead)
For center lines, section, dimension, extension, and phantom lines.

With a little practice, all lines mentioned above can be made with a 2H lead by varying pressure and relining.

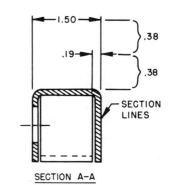

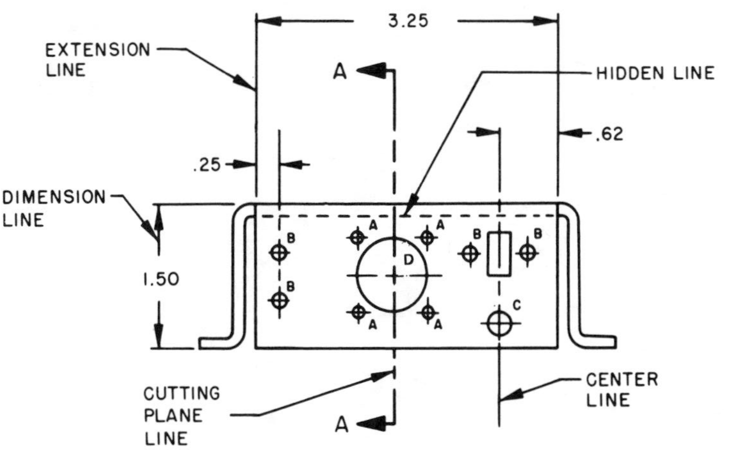

PANEL FRONT VIEW

Exercise. The panel above is only partially **dimensioned** at half scale. Redraw the panel full scale and dimension it completely. Omit section A-A.

Hint: The missing dimensions will have to be found by **scaling** the half scale drawing above and **multiplying by 2.**

All "A" holes should be dim. about the "D" hole center.
Note: Dim. lines may be extended to printed portions of this page.
Use your template for all holes and radii.

HOLE DIM. CHART	
LETTER	HOLE DIA.
A	.125
B	.188
C	.250
D	.750

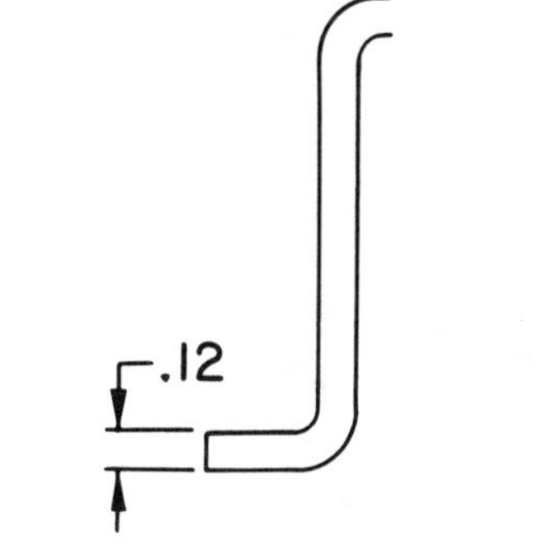

TITLE MECHANICAL DRAFTING REVIEW–DIMENSIONING, DOUBLE ARROW					DWG. NO. MDR-2	
NAME	DATE	COURSE	GRADE	SCALE NOTED	SHEET 1 OF 3	PAGE 3

DIMENSIONING FROM A DATUM

On sheet 1 you used the **conventional double arrow and dimension line** method, shown again in Fig. 1. If space is limited, **arrowless dimensioning from a datum** is preferred, as shown in Fig. 2.

Avoid dimensioning to hidden lines. Dimension another view. A **sectional view** may be necessary.

Use **decimal dimensioning** for all work.

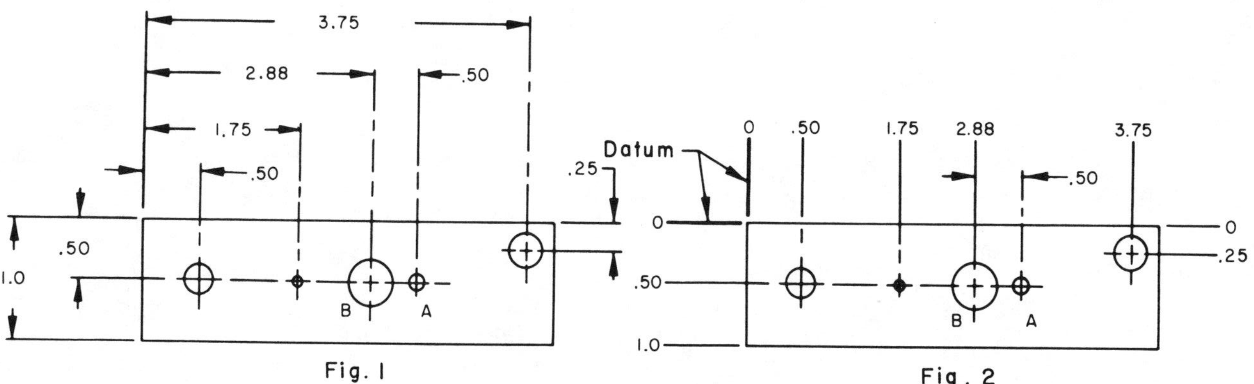

Fig. 1 Fig. 2

Note: Location of hole A is critical to hole B; therefore the dimension is direct to hole B and not to the datum.

Exercise. Redraw the front and section A-A of the panel on sheet 1 using the **datum dimensioning** method, full scale.

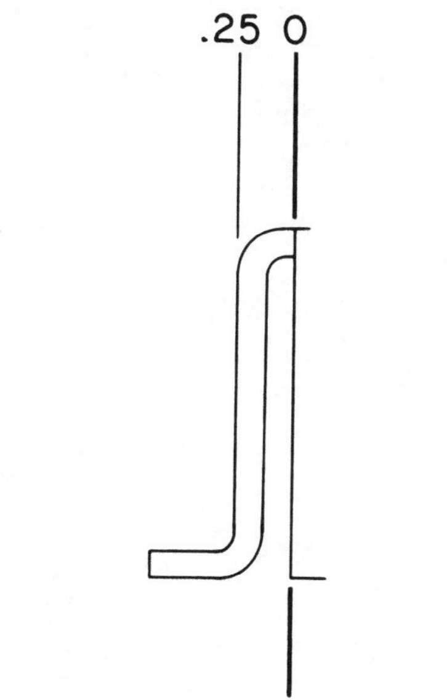

TITLE MECHANICAL DRAFTING REVIEW–DIMENSIONING, DATUM					DWG. NO. MDR-2	
NAME	DATE	COURSE	GRADE	SCALE NOTED	SHEET 2 OF 3	PAGE 4

DIMENSIONING

The drawing below is a typical partially dimensioned example showing the use of standard hole pattern layouts.

Exercise. Dimension the missing hole locations and draw the hole pattern properly in place. Use the **datum dimensioning method** (arrowless dimensioning). Scale: half size. Lettering ⅛″.

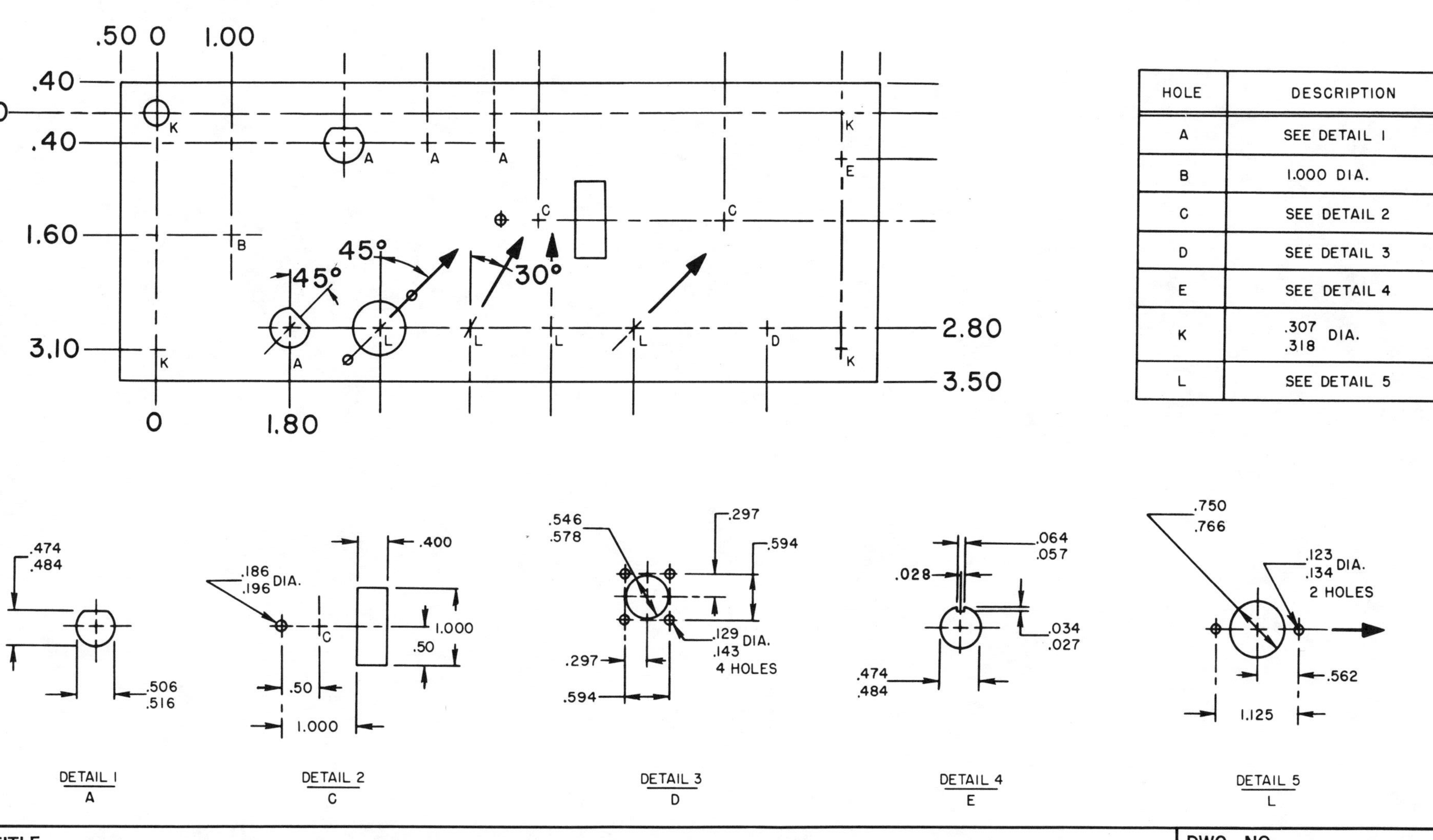

HOLE	DESCRIPTION
A	SEE DETAIL 1
B	1.000 DIA.
C	SEE DETAIL 2
D	SEE DETAIL 3
E	SEE DETAIL 4
K	.307 .318 DIA.
L	SEE DETAIL 5

TAPPED HOLES AND CLEARANCE HOLES

When a **part** has a feature, such as a tapped hole, which fixes a clamping screw or stud, the other part will have a **clearance hole** (see Fig. 1). Use the formulas in Appendix D, Single and Multiple Hole Mounting.

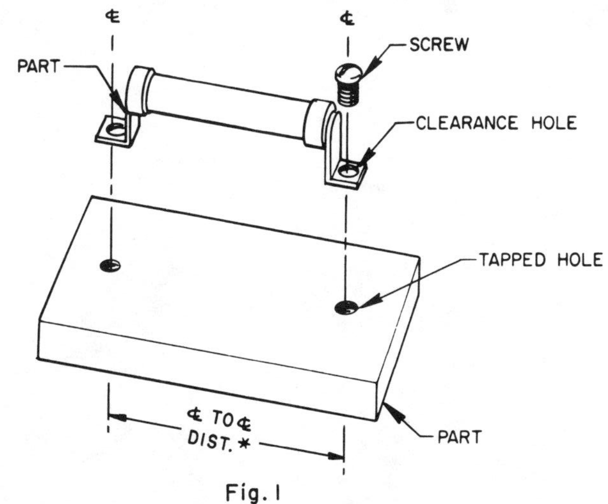

Fig. 1

Example. What is the **clearance hole diameter** of a single #10-32 binding head screw?

Answer. See Appendix D. Single Hole Dia. = .194. To this one must add the **tolerance,** which also is found in Appendix D under Standard Drilled Hole Tolerance. Since the hole is .194 Dia., look in the column opposite .126 through .250. The tolerance is $^{+.005}_{-.001}$ Therefore the correct way to specify the hole callout is $.194 ^{+.005}_{-.001}$ DIA.

For **multiple holes** use the formulas in Appendix D (see Multiple Hole Pattern). These formulas will be helpful in future work and design in the industry.

Study Appendix D examples 1, 2, and 3, then proceed to the exercise on the right.

*℄ = center line

Exercise. What is the correct clearance hole diameter and tolerance of these single binding head screws

#4-40 =

#¼-20 =

What is the correct clearance hole diameter, C'sink diameter, and tolerance of these flat head screws (82°)

#6-32 =

#10-32 =

Exercise. What should the **clearance hole diameter** and **tolerance** be in the part shown in Fig. 2? Show how you arrived at the answer. Use the formulas in Appendix D.

Answer

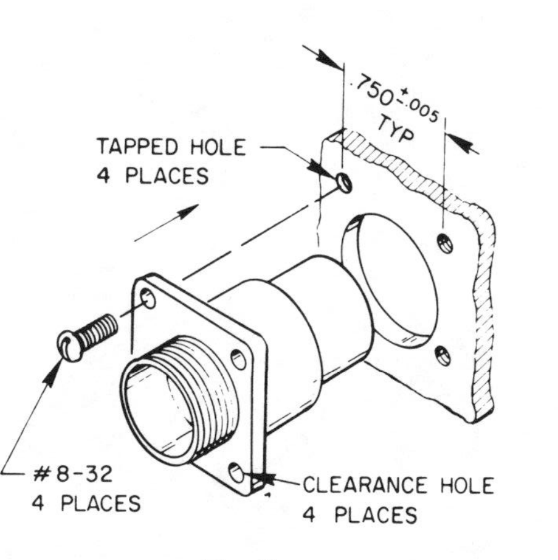

Fig. 2

TITLE **MECHANICAL DRAFTING REVIEW** – CLEARANCE HOLE ON TAPPED					DWG. NO. **MDR-3**	
NAME	DATE	COURSE	GRADE	SCALE NONE	SHEET 1 OF 2	PAGE 6

SCREW CLEARANCE HOLES

When a part is held by a **screw** and **nut** or fasteners which have the same basic clearance between **hole** and **screw** (see Fig. 1) use the same formulas shown in Appendix D. The same formulas will apply in this case *except that the last part is divided by 2.* Remember, the difference in formula depends on whether clearance holes (only) or tapped holes are being used.

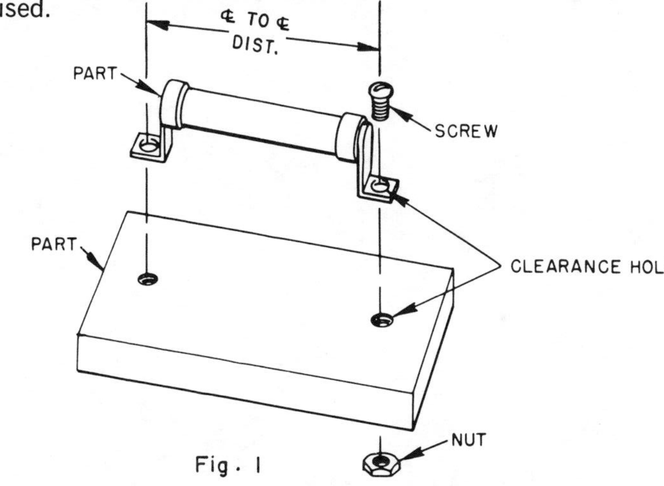

Fig. 1

Example. What should be the hole diameter of the sample shown in Fig. 1 if the two screws used are #6-32 binding head. (℄ to℄ is ±.010)?

The formula for the 2-hole (tapped) pattern in appendix D is: D = d + 2t. In this case (clearance hole on clearance hole) the last term is divided by 2 and reads:

$$D = d + \frac{2t}{2}$$

$$= .138 + \frac{2\,(.010)}{2}$$

$$= .138 + .010 = .148$$

The clearance hole will be (see drill sizes, Appendix E).

$$.149\ {}^{+.005}_{-.001}\ \text{DIA.}$$

2 HOLES

Exercise 1. What should be the clearance hole diameter and tolerance of the parts shown in Fig. 2?

Show how you arrived at the answer. Use the formulas in Appendix D.

Answer

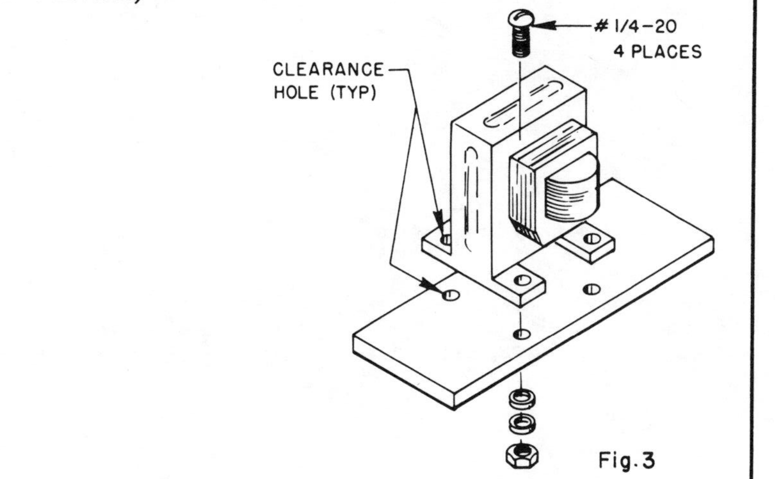

Fig. 2

Compare the answer above with previous exercise on sheet 1 (Fig. 2).

CLEARANCE ON TAPPED ________

CLEARANCE ON CLEARANCE ________

Exercise 2. What should be the clearance hole diameter and tolerance in Fig. 3? (℄ to℄ is ±.010)

Answer

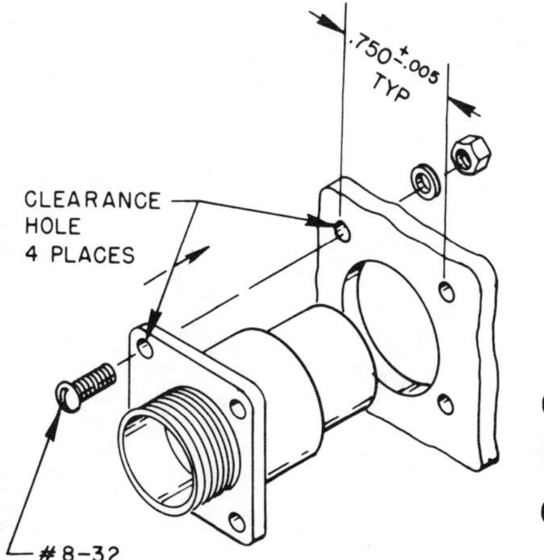

Fig. 3

TITLE MECHANICAL DRAFTING REVIEW-CLEARANCE HOLE ON CLEARANCE					DWG. NO. MDR-3	
NAME	DATE	COURSE	GRADE	SCALE NONE	SHEET 2 OF 2	PAGE 7

ELECTRO-GEOMETRIC CONSTRUCTIONS

Step-by-step **geometric construction technique** is shown in the accompanying tables and corresponding diagrams.

	GEOMETRIC CONSTRUCTION	SAMPLE USE
1	HEXAGON	HEX NUT
2	LINE DIVIDING	EQUAL SPACING OF HOLES
3	CONCENTRIC ARCS	SHEET METAL BENDS
4	LINE AND ARC TANGENTS	CABLE CLAMPS
5	TANGENT ARCS	POTENTIOMETER KNOB

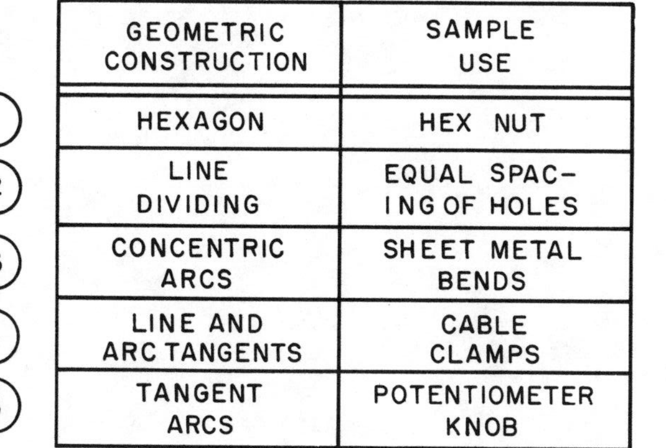

FIG. 1

Exercise. On the started illustration above (Fig. 1) lay out a **clamp** on a **clamp support.** Given:

R = 5/16, r = 1/8, L = 1.0 (**HINT:** For "r" corner,
W = 1/8, < = 30° see * of (3).)

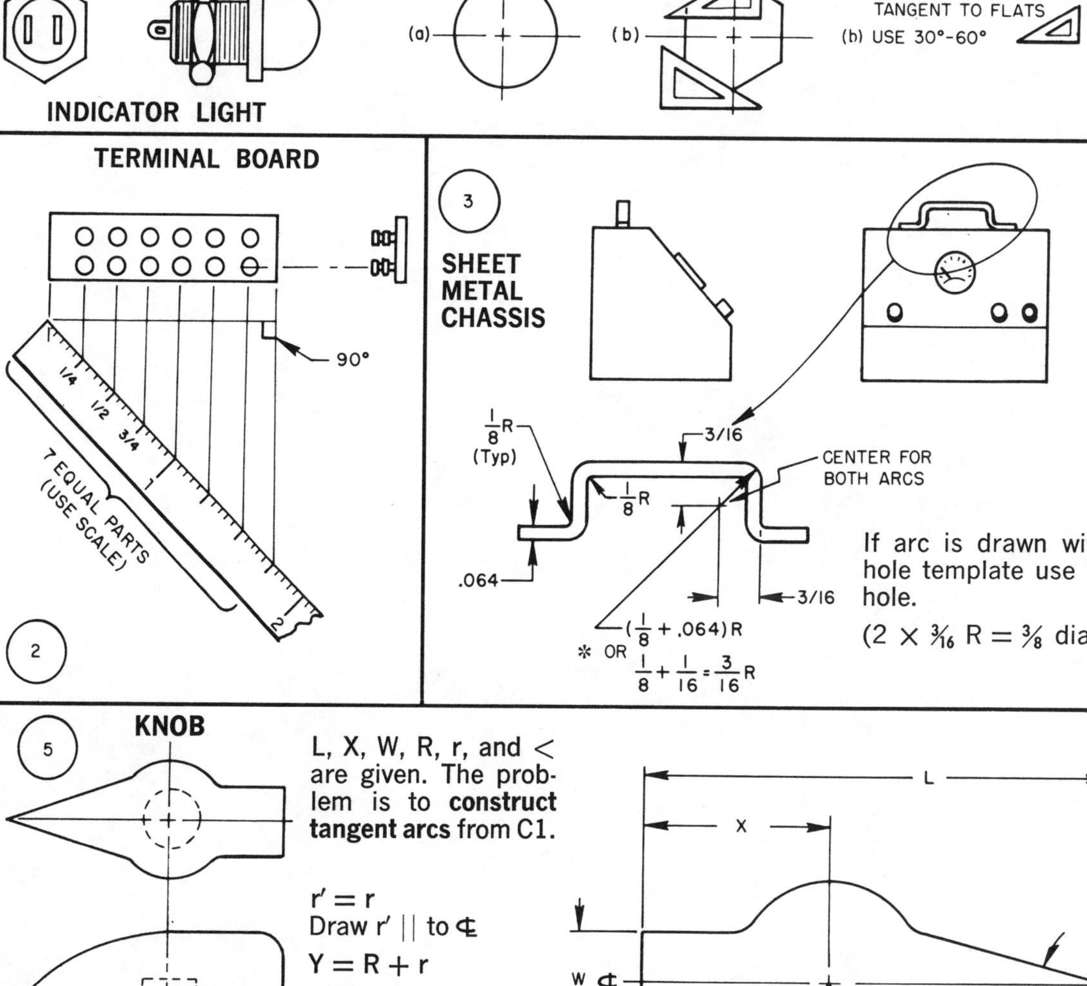

If arc is drawn with hole template use 3/8 hole.

(2 × 3/16 R = 3/8 dia.)

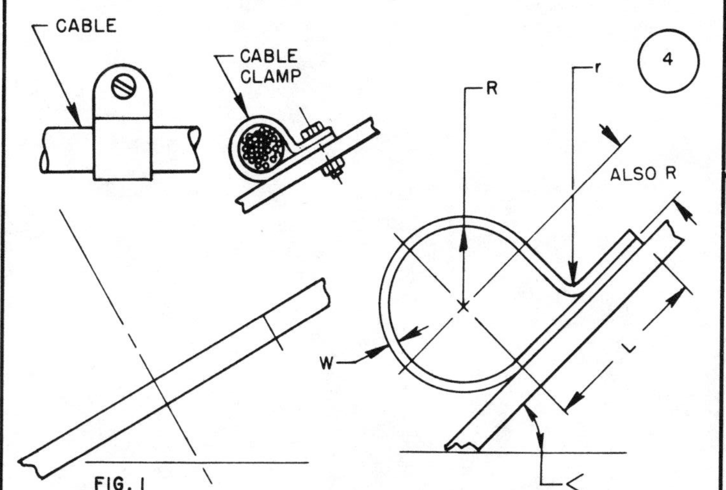

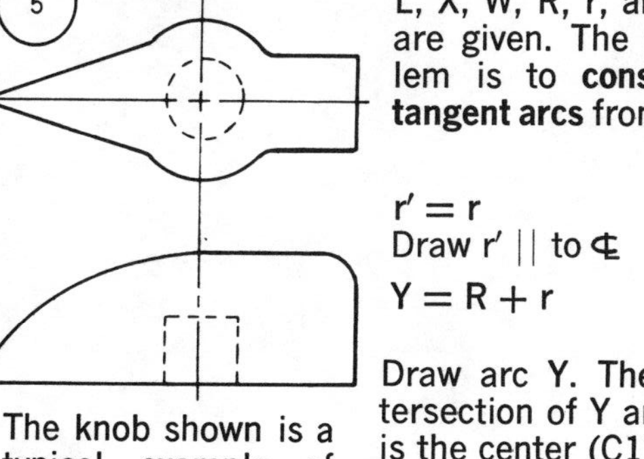

The knob shown is a typical example of **tangents and arcs** construction.

L, X, W, R, r, and < are given. The problem is to **construct tangent arcs** from C1.

$r' = r$
Draw r' || to ℄
$Y = R + r$

Draw arc Y. The intersection of Y and r' is the center (C1) for arc r.

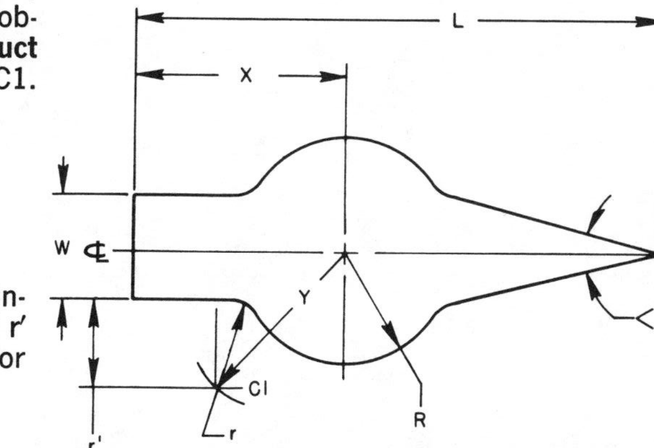

TITLE **MECHANICAL DRAFTING REVIEW**–GEOMETRIC CONSTRUCTION					DWG. NO. **MDR-4**	
NAME	DATE	COURSE	GRADE	SCALE	SHEET 1 OF 2	PAGE 8

Exercises. These are typical Electro-geometric construction exercises. Refer to sheet 1 for all exercises.

Draw a **hexagon** similar to the one shown on sheet 1, item 1, whose distance across the flats is 1.0″.

Draw a **terminal board** similar to the one shown on sheet 1, item 2, whose dimensions are 2 x 1.0 x $\frac{1}{16}$ with eight pairs of terminals forming two equally spaced rows.

Hint: Including edge distance, divide the board into nine equal parts.

Show your method of development by leaving in place all light construction lines.

See Appendix C (page 123) for terminal.

Draw a **knob** similar to the one shown on sheet 1, item 5, whose dimensions are:
$L = 2\frac{1}{2}$, $x = 1.0$, $R = \frac{1}{2}$
$W = \frac{1}{2}$, $r = \frac{3}{8}$, $\angle = 30°$
Construct the upper half of the knob by the method of **tangents and arcs,** then use your template to draw the lower half.

Draw a sheet-metal chassis **handle** similar to the one shown on sheet 1, item 3, whose dimensions are: $\frac{1}{8}$ thick, $\frac{3}{16}$ inside bend radius. The handle is $\frac{3}{4}$ high, 3.0 overall length with a $1\frac{3}{4}$ inside grip length. Construct the right half of the handle with concentric arcs and the left half with your template.

Draw a **cable clamp** similar to the one shown on sheet 1, item 4, for a cable 1.0″ diameter on a clamp support inclined 30°.
$W = \frac{1}{8}$, $L = 1\frac{1}{4}$, $r = \frac{1}{8}$.
Show construction lines.

TITLE MECHANICAL DRAFTING REVIEW–GEOMETRIC CONSTRUCTION					DWG. NO. MDR-4	
NAME	DATE	COURSE	GRADE	SCALE	SHEET 2 OF 2	PAGE 9

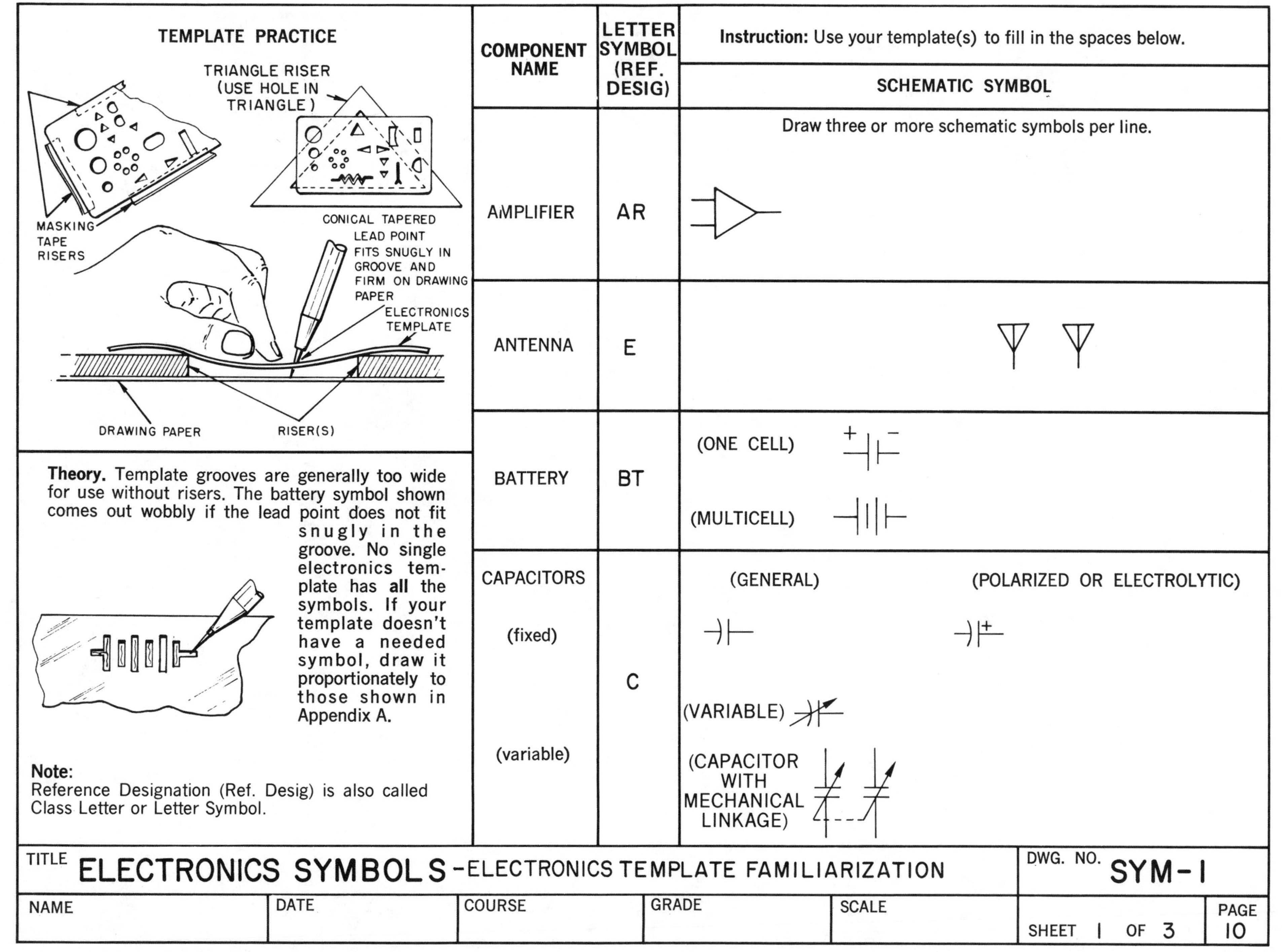

TEMPLATE PRACTICE

Theory. Template grooves are generally too wide for use without risers. The battery symbol shown comes out wobbly if the lead point does not fit snugly in the groove. No single electronics template has **all** the symbols. If your template doesn't have a needed symbol, draw it proportionately to those shown in Appendix A.

Note:
Reference Designation (Ref. Desig) is also called Class Letter or Letter Symbol.

Instruction: Use your template(s) to fill in the spaces below.

COMPONENT NAME	LETTER SYMBOL (REF. DESIG)	SCHEMATIC SYMBOL
		Draw three or more schematic symbols per line.
AMPLIFIER	AR	
ANTENNA	E	
BATTERY	BT	(ONE CELL) + − (MULTICELL)
CAPACITORS (fixed) (variable)	C	(GENERAL) (POLARIZED OR ELECTROLYTIC) + (VARIABLE) (CAPACITOR WITH MECHANICAL LINKAGE)

TITLE **ELECTRONICS SYMBOLS** - ELECTRONICS TEMPLATE FAMILIARIZATION

NAME	DATE	COURSE	GRADE	SCALE	DWG. NO. SYM-1 SHEET 1 OF 3	PAGE 10

Fill in the blank spaces as in the last exercise by referring to Appendix A. Use either vertical or inclined 1/8″ or 1/10″ uppercase lettering (three symbols per line).

COMPONENT NAME	REF DESIG	SCHEMATIC SYMBOL	COMPONENT NAME	REF DESIG	SCHEMATIC SYMBOL
SEMICONDUCTOR, RECTIFIER DIODE					
	Y		INCANDESCENT FILAMENT		
	TB			LS	
		A B C D	5-CONDUCTOR CABLE		

TITLE ELECTRONICS SYMBOLS-ELECTRONICS TEMPLATE FAMILIARIZATION

DWG. NO. SYM-1

NAME	DATE	COURSE	GRADE	SCALE	SHEET 2 OF 3	PAGE 11

Choose nine new symbols from Appendix A which were not drawn in the last two exercises and complete the chart below. Using all the space available draw three symbols per line symmetrically, with $\frac{1}{8}$" lettering (lightly make your layout first).

COMPONENT NAME	REF. DESIG.	SCHEMATIC SYMBOLS

TITLE **ELECTRONICS SYMBOLS**–ELECTRONICS TEMPLATE FAMILIARIZATION

DWG. NO. SYM-1

NAME	DATE	COURSE	GRADE	SCALE	SHEET 3 OF 3	PAGE 12

SYMBOL HOOKUP PRACTICE

Study the printed circuit (PC) hookups shown below with their corresponding symbol representations (circuits). Complete the circuits in columns 2 and 3.

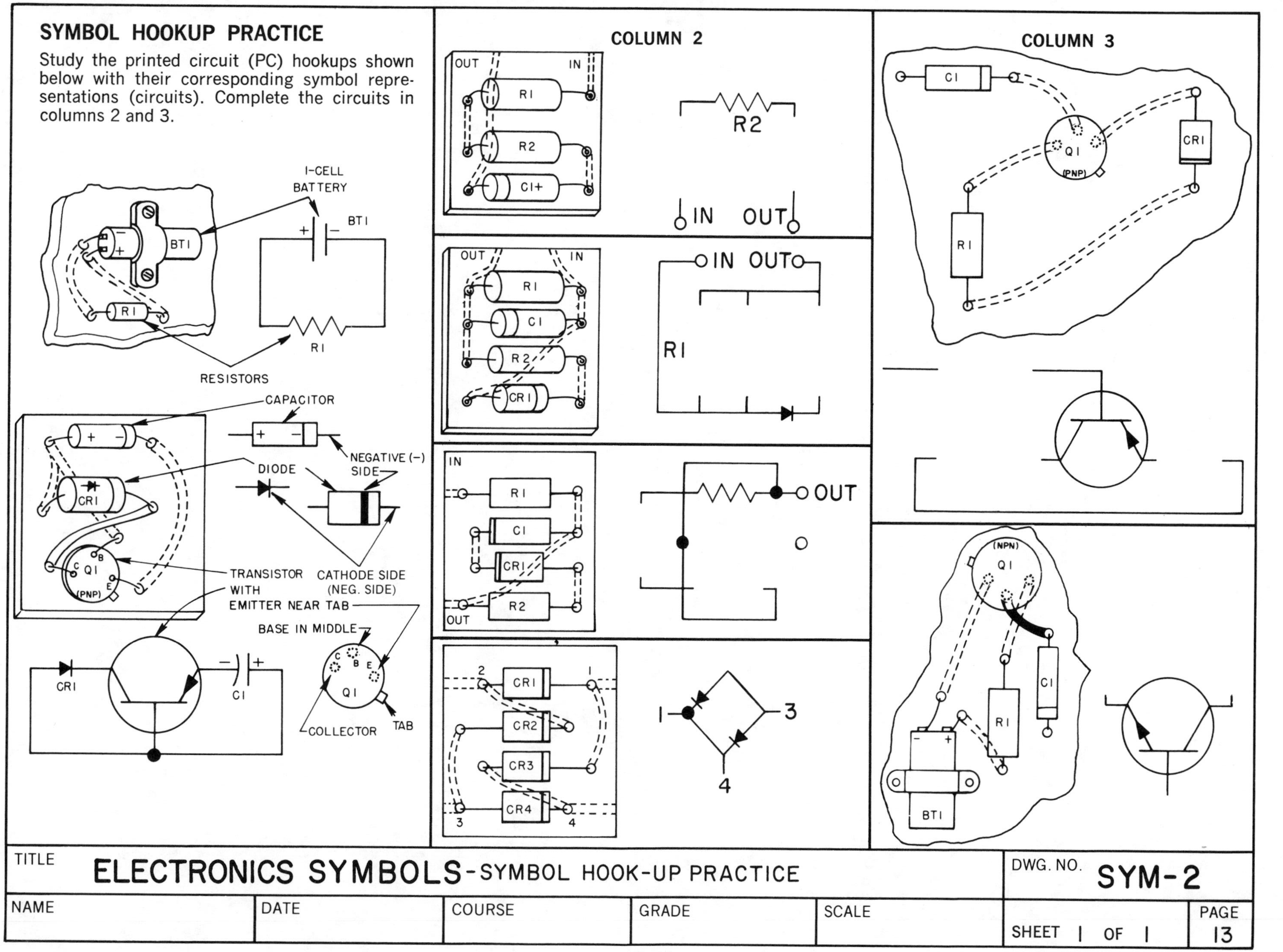

TITLE **ELECTRONICS SYMBOLS**-SYMBOL HOOK-UP PRACTICE

DWG. NO. **SYM-2**

NAME	DATE	COURSE	GRADE	SCALE	SHEET 1 OF 1	PAGE 13

INTRODUCTION: COMPONENT OUTLINE DRAWINGS

An **outline drawing,** as shown below, describes the contour and overall dimensions of a component. These drawings are usually associated with "specification control" or "purchase control" drawings (the three names are often used interchangeably). The "spec" control drawing consists of an outline drawing plus other information which specifies the necessary prerequisites for use of the component. The drawing generally includes manufacturer's name, model number, component name and symbol plus a list of specifications. This type of drawing is regarded as the principal document in a component purchasing contract.

EXAMPLE

COMPONENT OUTLINE

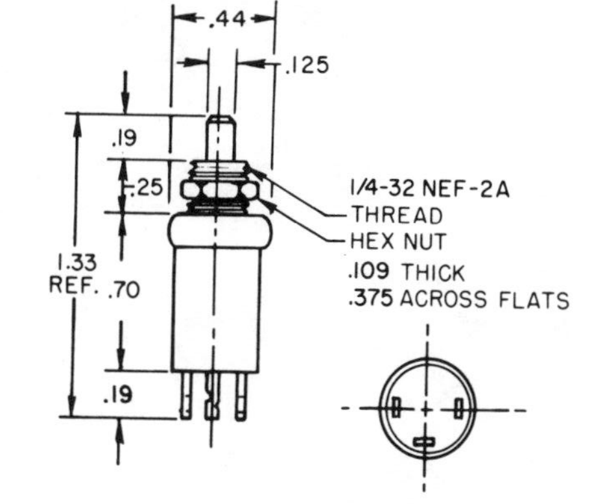

SINGLE POLE, DOUBLE THROW SWITCH

RATING:

Rated: to make and break ¼ ampere, 115 VAC resistive.
Contact Resistance: .010 ohms maximum initial measured at 2 VDC, 100 ma. After 250,000 operations .010 ohms typical; .020 ohms maximum.
Insulation Resistance: 100,000 megohms measured at 100 VDC, 60% to 70% relative humidity.
Dielectric Strength: 1500 VAC approximate at sea level.
Life Expectancy: 250,000 operations at rated load.

CONSTRUCTION:

Single pole double throw
Wiping contacts
Button Travel: .187″ approximately.
Overtravel: .062″ approximately.
Actuating Force: 16 oz. approximately to bottom button.
Mounting Hole: 17/64″ dia.

COMPONENT SYMBOL

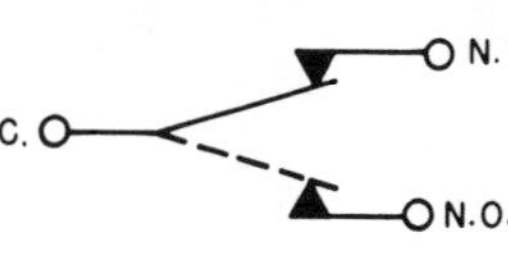

MANUFACTURER:
GRAYHILL INC.

Exercise. Draw an **outline** of a ½AMP 125V fuse, full size. Show all dimensions, symbol, reference designation, manufacturer's name, and component part number. Information may be found in Appendix B, page 120.

COMPONENT
OUTLINE

COMPONENT SYMBOL ∿ ____________

REF. DESIGNATION ____________

MANUFACTURER ____________

PART NO. ____________

Exercise. Give the following Mil Type Designation No. of 1.2K, 2.4K, 15K, 56K, 390K, 6.2MEG. ½WATT ±5% resistor (RC20 type). Information may be found in Appendix B, page 116.

RESISTANCE IN OHMS	MIL TYPE DESIGNATION
1.2K (1200)	RC20GF122J

TITLE COMPONENT OUTLINE - INTRODUCTION					DWG. NO. COMP-1	
NAME	DATE	COURSE	GRADE	SCALE NOTED	SHEET 1 OF 2	PAGE 14

SPECIFICATION CONTROL DRAWING

Complete the exercises below. All outline drawing information may be found in Appendix B.

Make an outline drawing of transistor 2N1304, 2 × size. Show (two views), all dimensions, symbol, and Manufacturer's name.

COMPONENT SYMBOL

NPN

MANUFACTURER	REF. DESIG.	COMPONENT
	Q	TRANSISTOR 2N1304

Make an outline drawing of 26.570 MC. crystal, 2 × size. Show (two views) all dimensions, symbol, reference designation, Manufacturer's name, and component number.

MANUFACTURER	REF. DESIG.	COMPONENT
		CRYSTAL

TITLE	COMPONENT OUTLINE–SPECIFICATION CONTROL DRAWINGS				DWG. NO. COMP-1	
NAME	DATE	COURSE	GRADE	SCALE	SHEET 2 OF 2	PAGE 15

A TYPICAL PROBLEM IN OUTLINE DRAWING. An engineer has given you the following list of components to be used in a packaging design:

A. RESISTOR, FIXED, COMPOSITION, ¼WATT ±5% (MIL TYPE RC07)
R1, R3: 1K; R2, R5, R7: 33K; R4, R6, R8: 22K; R9, R11: 100K; R12, R10: 1.2MEG

B. CAPACITOR, GENERAL PURPOSE, 200V (MIL TYPE CK05)
C1, C5, C7: 47pf; C2, C4, C6: 120pf; C3: 150pf; C8, C10: 470pf; C9, C11: 820pf (Use 20% tolerance)

C. CAPACITOR, ELECTROLYTE, 35V, ±10% (MIL TYPE CS13)
C12, C14: 0.33MFD; C13, C15: 0.47MFD; C16, C17: 1MFD

Prepare three types of "spec control" drawings (A, B, and C) in the three columns below. The first column is almost complete, the second is just started, and the third is blank. Complete all three columns in a similar manner, using Appendix B. All drawings 2 × size.

A — RESISTOR — RC07

REF. DESIG.	RESISTANCE IN OHMS	MIL TYPE DESIGNATION
R1, R3	1000	RC07GF102J
R2, R5, R7	33K	
R4, R6, R8		

COMPONENT OUTLINE
(COMPLETE DWG. BELOW)

COMPONENT SYMBOL
(DRAW IT BELOW)

B — CAPACITOR — CK05

REF. DESIG.	CAPACITANCE IN pf	MIL TYPE DESIGNATION

COMPONENT OUTLINE

COMPONENT SYMBOL

MANUFACTURER

C — CAPACITOR — CS13

TITLE					DWG. NO.	
COMPONENT OUTLINE—ELECTRICAL					COMP-2	
NAME	DATE	COURSE	GRADE	SCALE 2 X SIZE	SHEET 1 OF 2	PAGE 16

Example. The block diagram shown below explains the pictorial diagram of Fig. 1, in which a microphone is hooked to the input of the amplifier and the output of the amplifier is hooked to the speaker. Notice direction of arrowheads from input to output.

BLOCK DIAGRAM

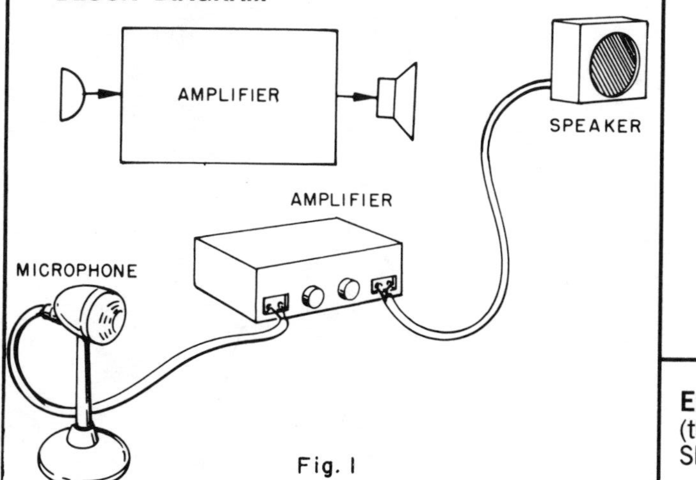

Fig. 1

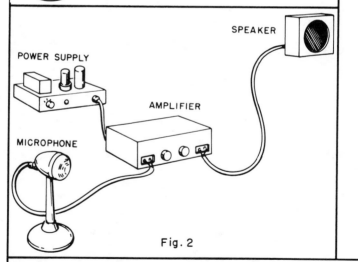

POWER SUPPLY

SPEAKER

AMPLIFIER

MICROPHONE

Fig. 2

Exercise 1. An identical setup of a simple intercom is shown in Fig. 2. The only difference is that a **power supply** was added to the **amplifier.** Draw a block diagram of the intercom of Fig. 2. Show the correct direction of arrowheads. (**Hint:** Power supplies are shown at bottom)

Exercise 2. Draw the same block diagram as you have shown in exercise 1, but add three more speakers (that is, four speakers total) hooked in parallel from the one output of the amplifier shown in Fig. 2. Show all arrowheads.

TITLE **BLOCK DIAGRAM** - INTERCOM					DWG. NO. **BD-2**	
NAME	DATE	COURSE	GRADE	SCALE	SHEET 1 OF 1	PAGE 19

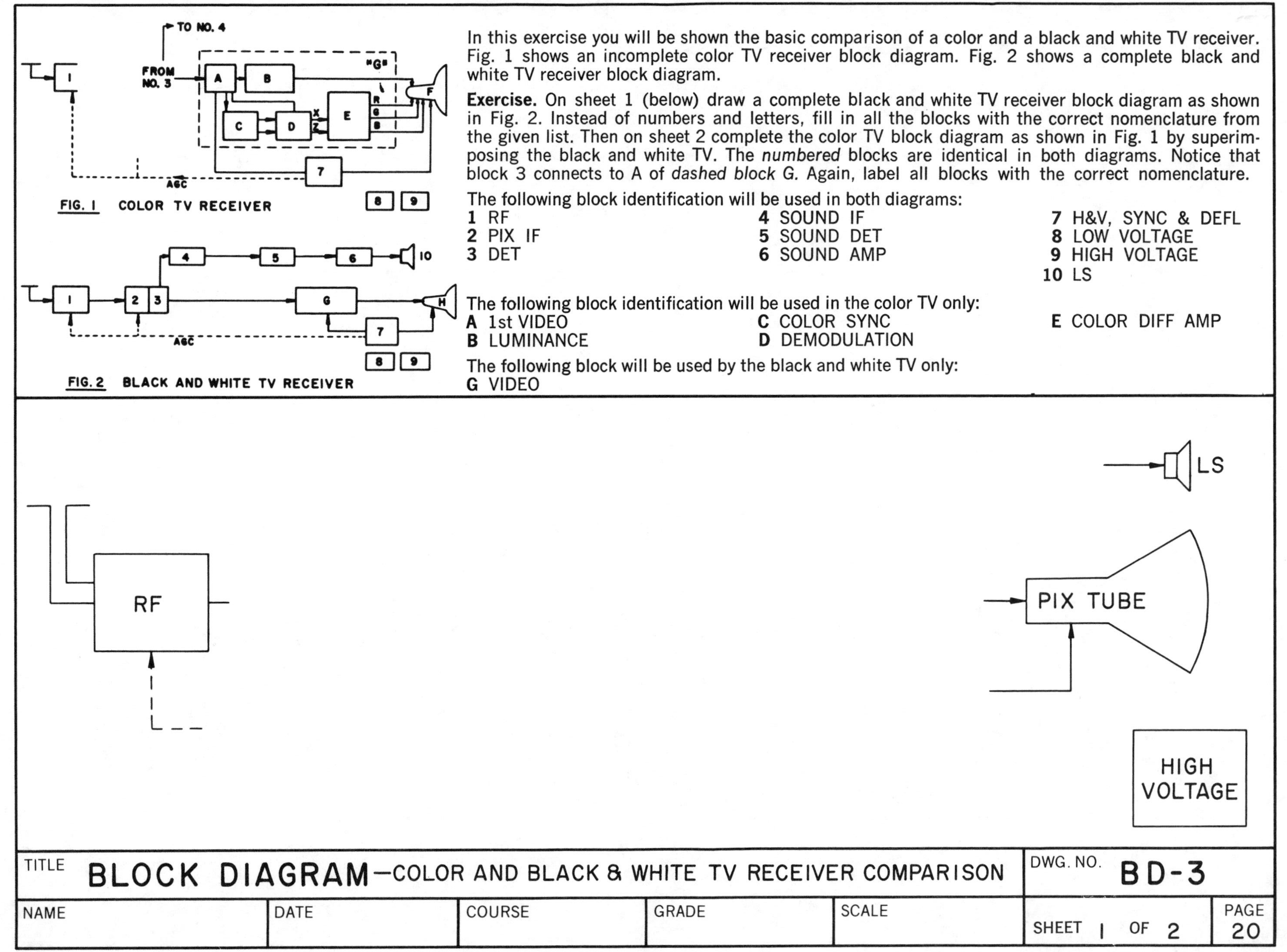

In this exercise you will be shown the basic comparison of a color and a black and white TV receiver. Fig. 1 shows an incomplete color TV receiver block diagram. Fig. 2 shows a complete black and white TV receiver block diagram.

Exercise. On sheet 1 (below) draw a complete black and white TV receiver block diagram as shown in Fig. 2. Instead of numbers and letters, fill in all the blocks with the correct nomenclature from the given list. Then on sheet 2 complete the color TV block diagram as shown in Fig. 1 by superimposing the black and white TV. The *numbered* blocks are identical in both diagrams. Notice that block 3 connects to A of *dashed block G.* Again, label all blocks with the correct nomenclature.

The following block identification will be used in both diagrams:

1 RF
2 PIX IF
3 DET
4 SOUND IF
5 SOUND DET
6 SOUND AMP
7 H&V, SYNC & DEFL
8 LOW VOLTAGE
9 HIGH VOLTAGE
10 LS

The following block identification will be used in the color TV only:

A 1st VIDEO
B LUMINANCE
C COLOR SYNC
D DEMODULATION
E COLOR DIFF AMP

The following block will be used by the black and white TV only:

G VIDEO

TITLE BLOCK DIAGRAM—COLOR AND BLACK & WHITE TV RECEIVER COMPARISON					DWG. NO. BD-3	
NAME	DATE	COURSE	GRADE	SCALE	SHEET 1 OF 2	PAGE 20

Complete the started block diagram of the color TV receiver below, using sheet 1 (page 20) and list of blocks for reference. Label all blocks with the correct nomenclature.

LS

RF

PIX TUBE

HIGH VOLTAGE

TITLE BLOCK DIAGRAM–COLOR AND BLACK & WHITE TV RECEIVER COMPARISON					DWG. NO. BD-3	
NAME	DATE	COURSE	GRADE	SCALE	SHEET 2 OF 2	PAGE 21

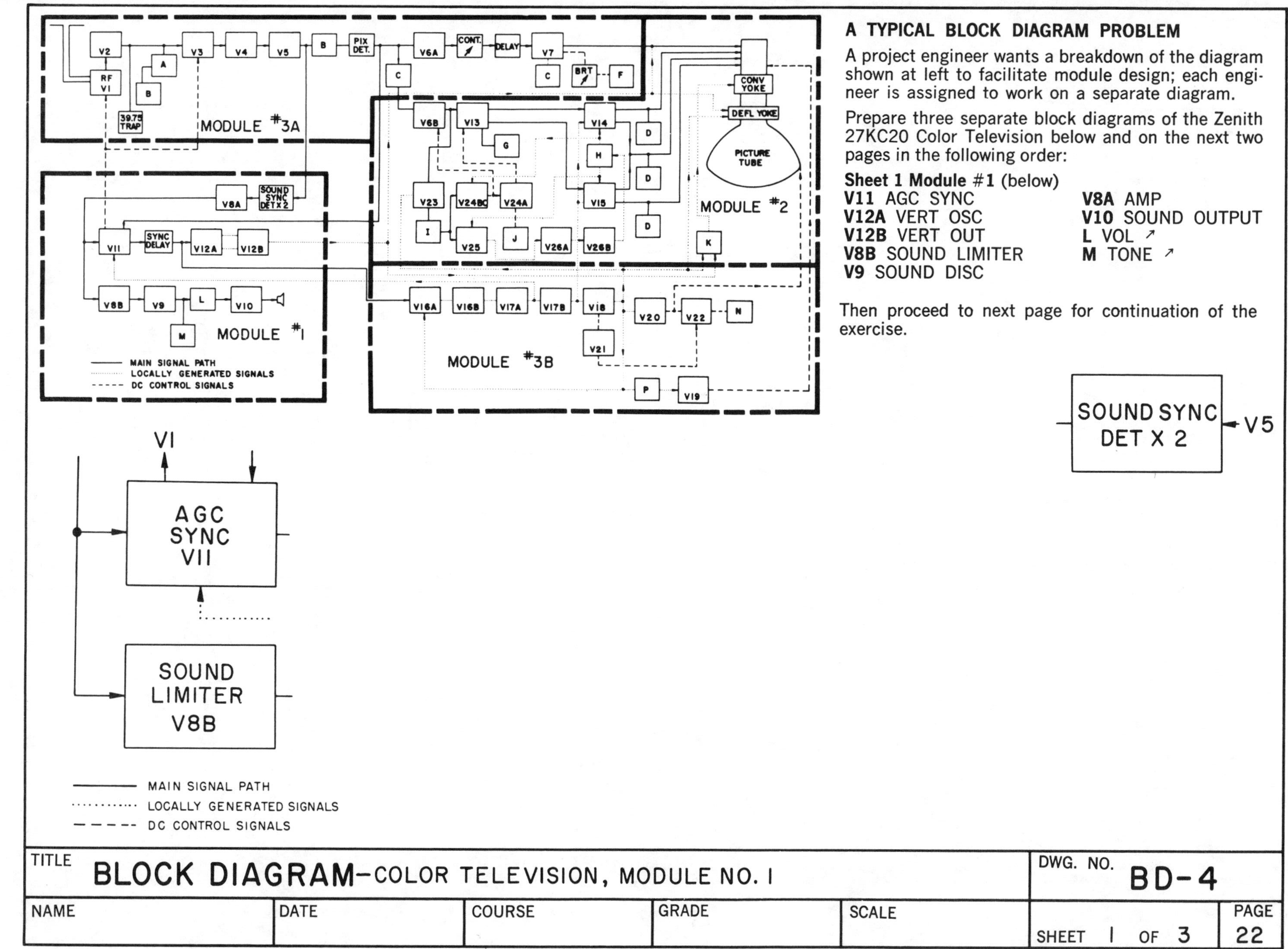

A TYPICAL BLOCK DIAGRAM PROBLEM

A project engineer wants a breakdown of the diagram shown at left to facilitate module design; each engineer is assigned to work on a separate diagram.

Prepare three separate block diagrams of the Zenith 27KC20 Color Television below and on the next two pages in the following order:

Sheet 1 Module #1 (below)

V11 AGC SYNC
V12A VERT OSC
V12B VERT OUT
V8B SOUND LIMITER
V9 SOUND DISC
V8A AMP
V10 SOUND OUTPUT
L VOL ↗
M TONE ↗

Then proceed to next page for continuation of the exercise.

TITLE BLOCK DIAGRAM—COLOR TELEVISION, MODULE NO. I					DWG. NO. BD-4	
NAME	DATE	COURSE	GRADE	SCALE	SHEET I OF 3	PAGE 22

Complete the color television block diagram of Module #2 below. Use sheet 1 (page 22) for reference, use your own judgment for block spacing and lettering.

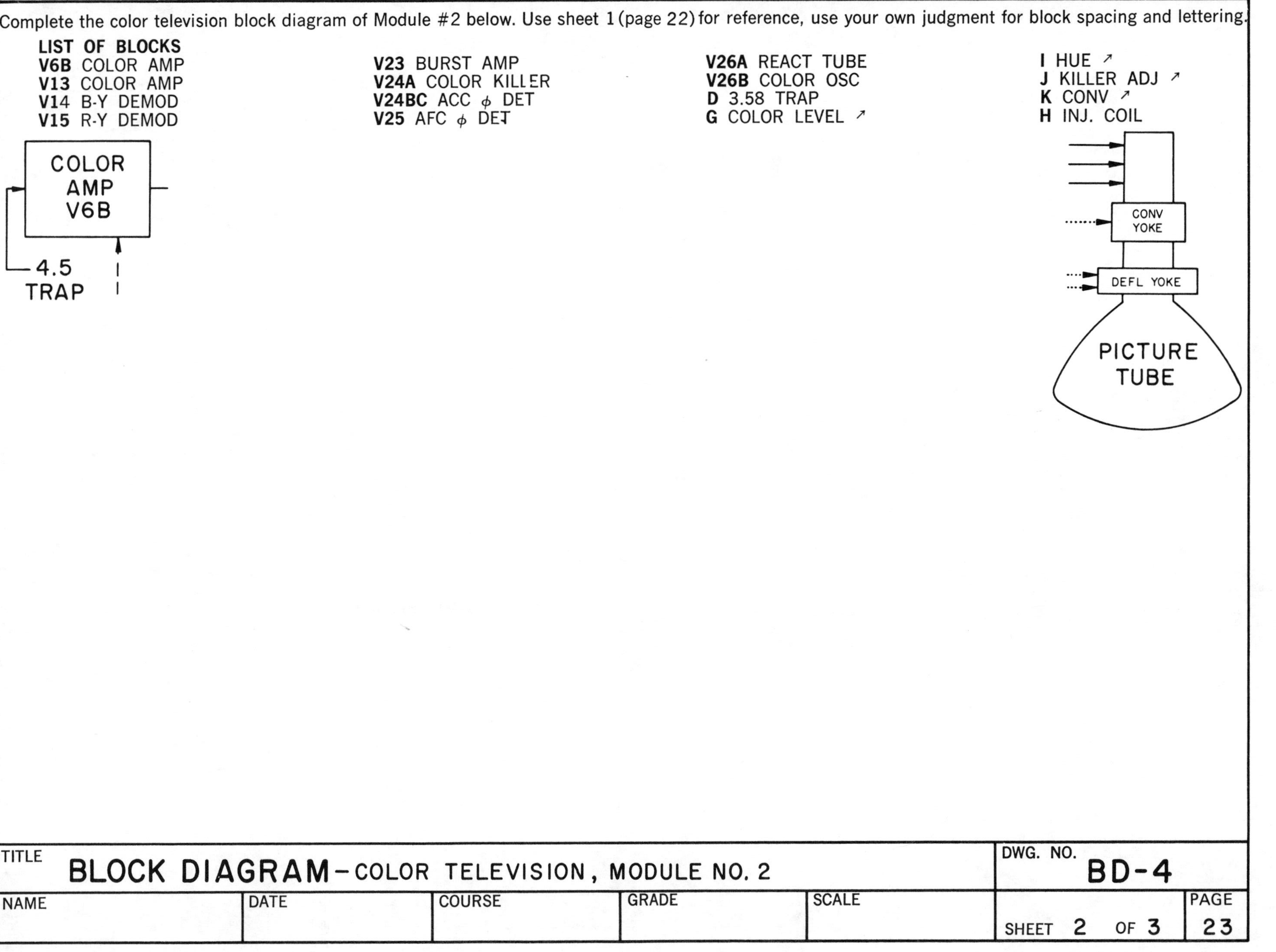

TITLE BLOCK DIAGRAM – COLOR TELEVISION, MODULE NO. 2					DWG. NO. BD-4	
NAME	DATE	COURSE	GRADE	SCALE	SHEET 2 OF 3	PAGE 23

Complete the color television block diagram of Modules #3A and #3B. Use sheet 1 (page 22) for reference, use your own judgment for block spacing and lettering. Orientation of blocks does not have to be exactly as shown in sheet 1 as long as the flow (direction) is correct.

LIST OF BLOCKS

V2 CONV	**V5** 3rd IF	**V16A** HORIZ ϕ DET	**V17B** HORIZ DISCH	**V20** HV RECT	**A** 47.25 TRAP	**F** BRT RANGE ↗
V3 1st IF	**V6A** CATH FOLL	**V16B** HORIZ CONTR	**V18** HORIZ OUT	**V21** DAMPER	**B** 41.25 TRAP	**N** HV ↗
V4 2nd IF	**V7** Y AMP	**V17A** HORIZ OSC	**V19** FOCUS RECT	**V22** HV REG	**C** 4.5 TRAP	**P** FOCUS ↗

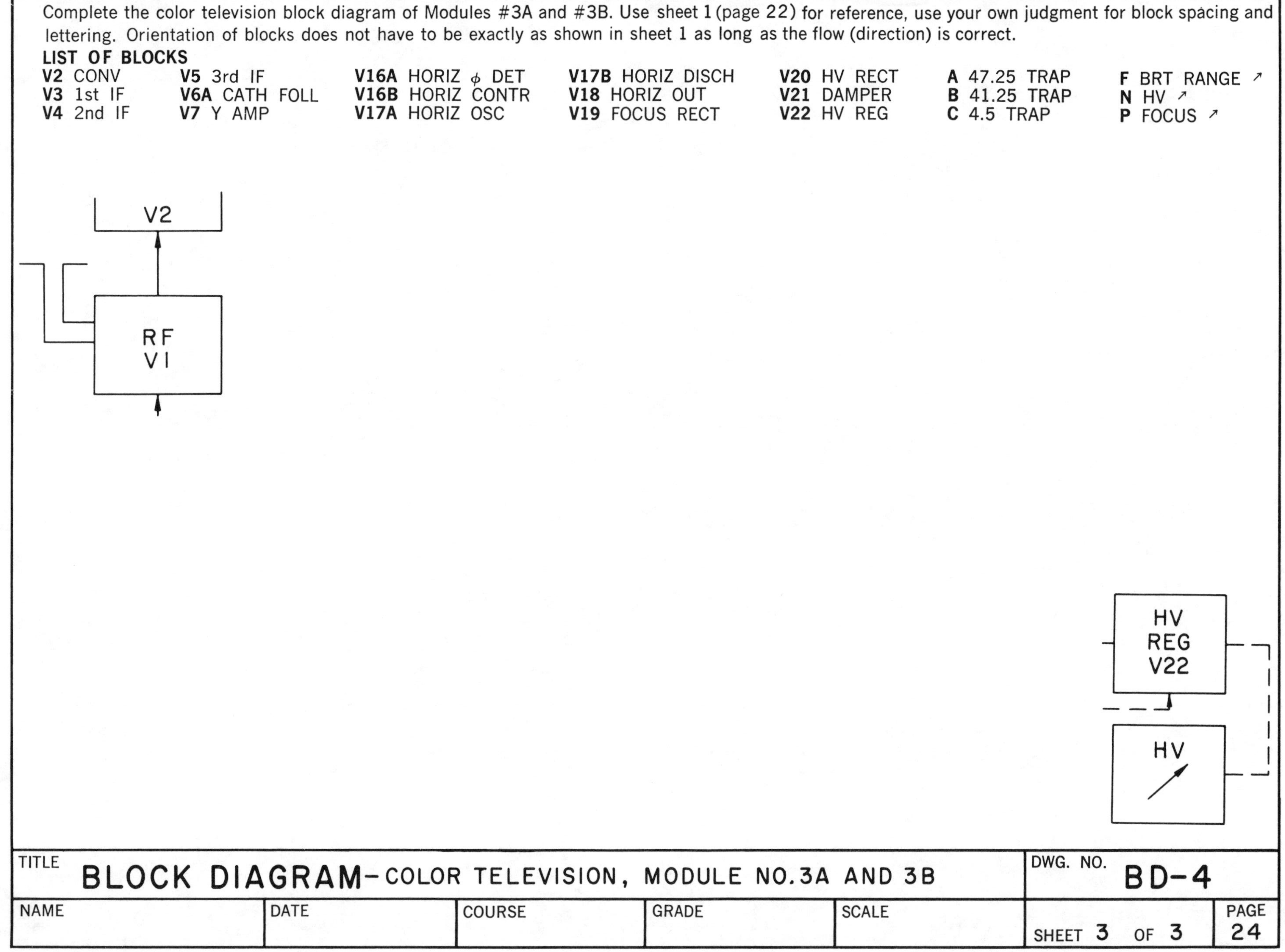

TITLE **BLOCK DIAGRAM**–COLOR TELEVISION, MODULE NO. 3A AND 3B

DWG. NO. **BD-4**

NAME	DATE	COURSE	GRADE	SCALE	SHEET 3 OF 3	PAGE 24

INTRODUCTION: THE SCHEMATIC DIAGRAM

The best way to learn how to draw a schematic diagram is to follow a good example (like SCH-3 on page 27 of this lesson, the **FM Tuner**). However, a few other tips are worth mentioning.

1. Generally, lay out the spacing of transistors (or tubes) first and surrounding circuitry second.
2. Be sure to leave enough space for lettering between symbols.
3. Draw the complete layout lightly first and "heavy up" afterwards.

At right are properly drawn wire junctions, crossings and typical symbol identifications and ratings. SCH-3, illustrates these and the lettering of notes. Notice that the last numbers used are included with the notes. Omitted numbers should also be listed as omitted (this often occurs after revisions have been made).

TUBES: A typical tube symbol is shown in Fig. 1, however, a simpler and preferred diagram can be drawn as shown in Fig. 2.

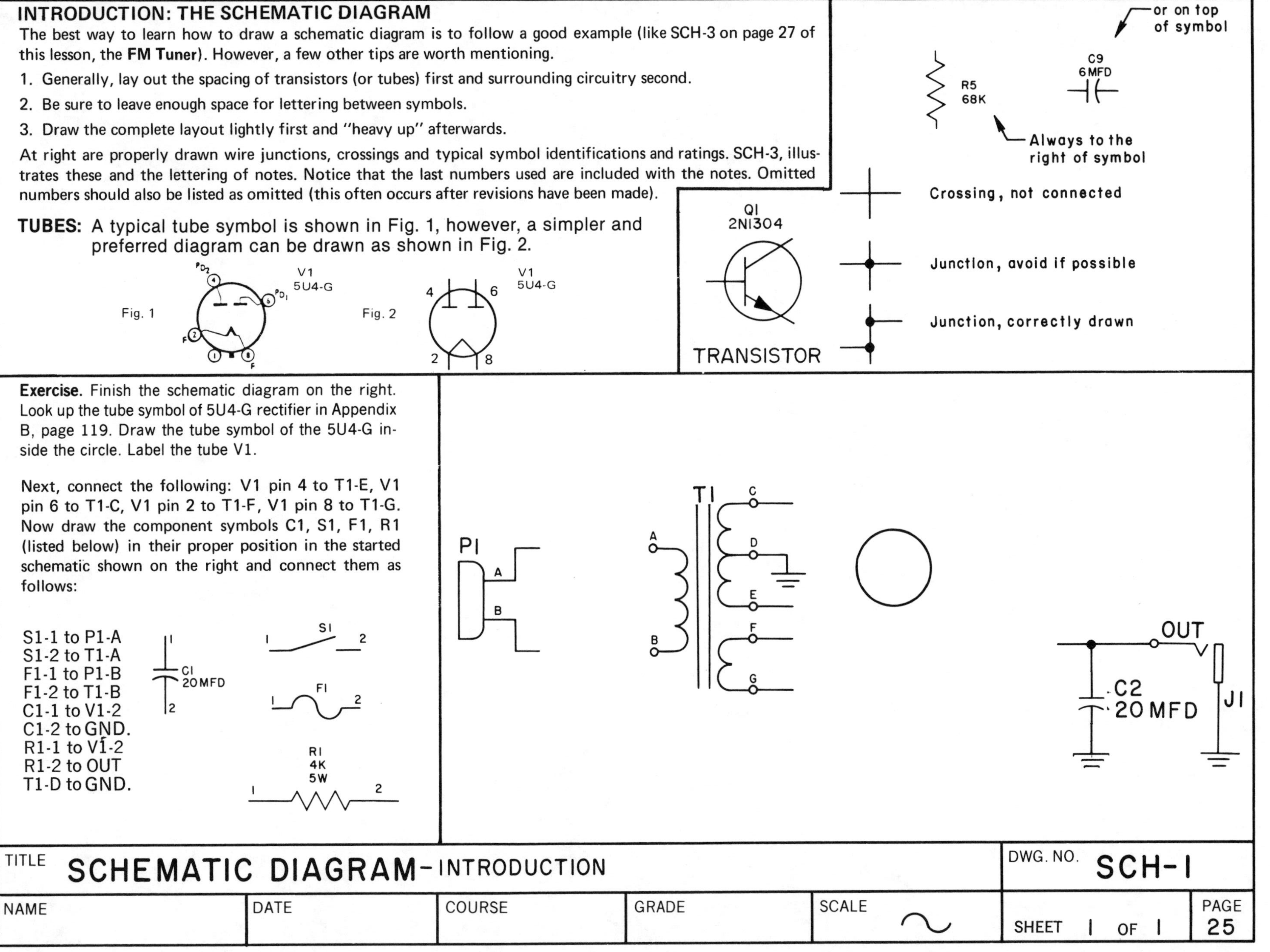

Exercise. Finish the schematic diagram on the right. Look up the tube symbol of 5U4-G rectifier in Appendix B, page 119. Draw the tube symbol of the 5U4-G inside the circle. Label the tube V1.

Next, connect the following: V1 pin 4 to T1-E, V1 pin 6 to T1-C, V1 pin 2 to T1-F, V1 pin 8 to T1-G. Now draw the component symbols C1, S1, F1, R1 (listed below) in their proper position in the started schematic shown on the right and connect them as follows:

S1-1 to P1-A
S1-2 to T1-A
F1-1 to P1-B
F1-2 to T1-B
C1-1 to V1-2
C1-2 to GND.
R1-1 to V1-2
R1-2 to OUT
T1-D to GND.

TITLE SCHEMATIC DIAGRAM-INTRODUCTION					DWG. NO. SCH-1	
NAME	DATE	COURSE	GRADE	SCALE ~	SHEET 1 OF 1	PAGE 25

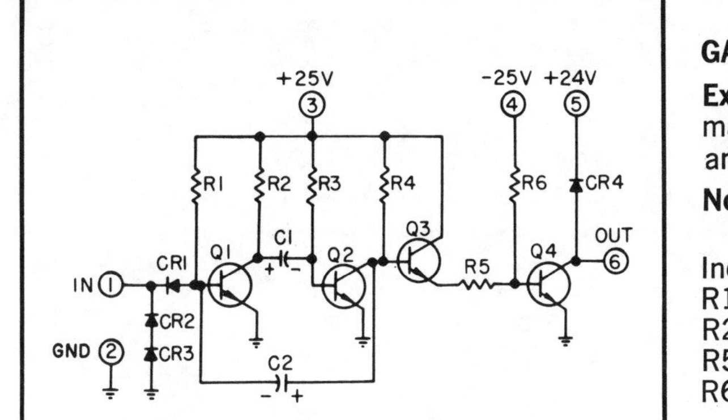

GATED MULTIVIBRATOR, SCHEMATIC DIAGRAM

Exercise. From the schematic diagram shown at left, complete the schematic below according to the ASA manner (i.e., show identification and rating of each component). Use your own electronics template and complete the notes in the drawing below as follows:

Notes: Unless otherwise specified, 1. all resistance values are in ohms ¼ watt ±5%.

Indicate component ratings from the given table:

R1 and R3 = 560K
R2 and R4 = 22K
R5 = 33K
R6 = 470K

All capacitors = 2.2MFD, 35V
All diodes = 1N617
All transistors = 2N1304

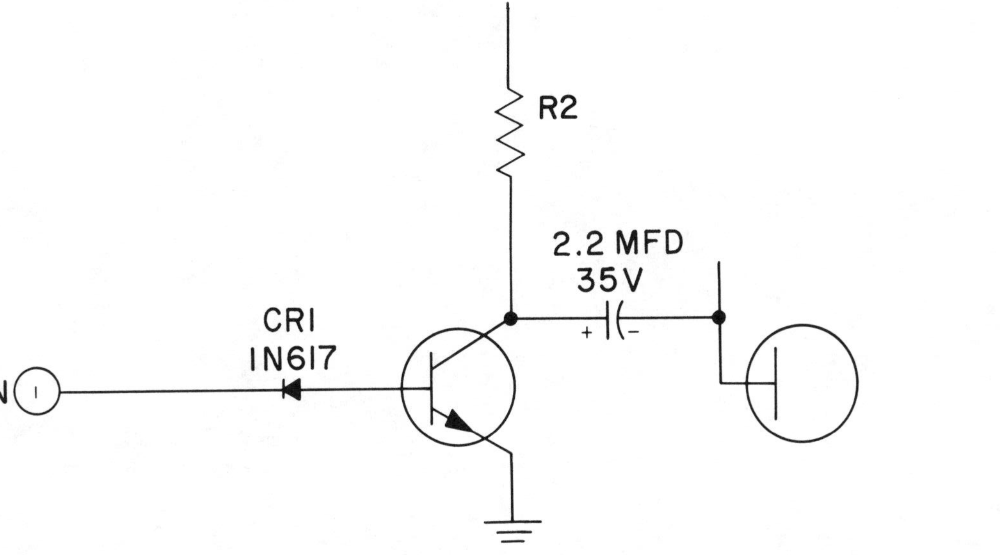

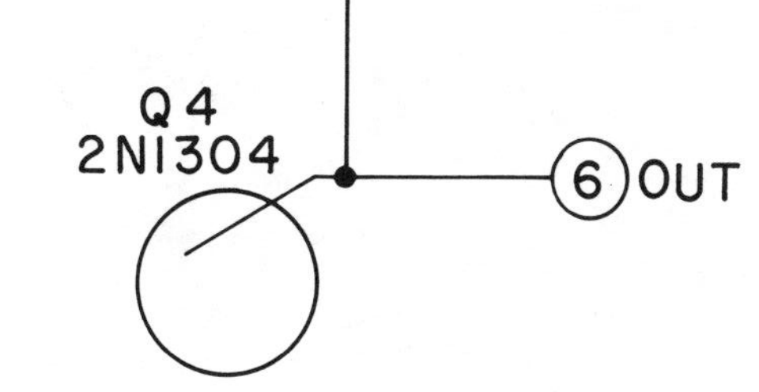

NOTES: UNLESS OTHERWISE

1.

TITLE SCHEMATIC DIAGRAM—GATED MULTIVIBRATOR					DWG. NO. SCH-2	
NAME	DATE	COURSE	GRADE	SCALE ∼	SHEET 1 OF 1	PAGE 26

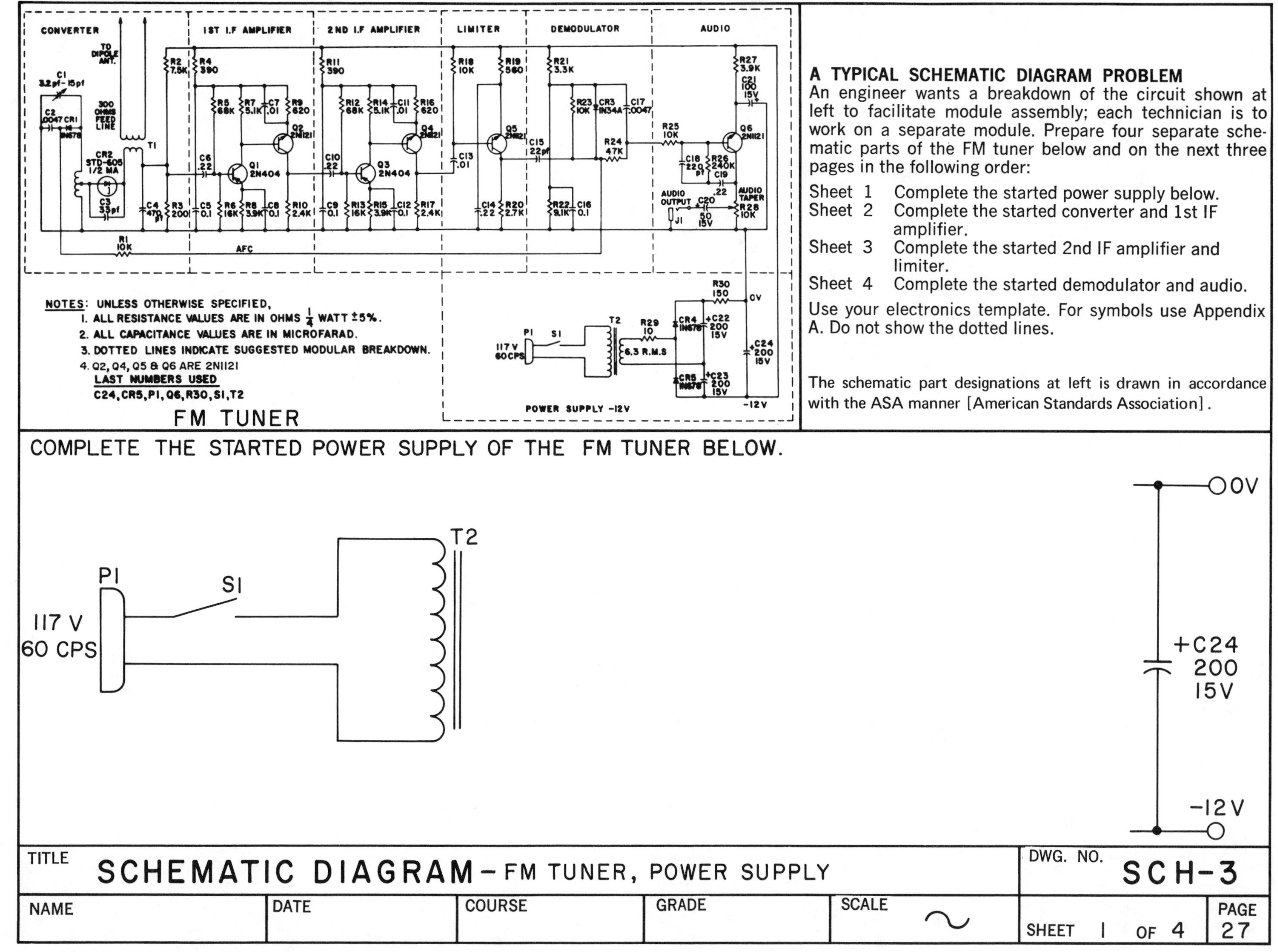
CONVERTER
1ST I.F AMPLIFIER
2ND I.F AMPLIFIER
LIMITER
DEMODULATOR
AUDIO
TO DIPOLE ANT.
300 OHMS FEED LINE
AFC
AUDIO OUTPUT
AUDIO TAPER
POWER SUPPLY -12V
NOTES: UNLESS OTHERWISE SPECIFIED,
1. ALL RESISTANCE VALUES ARE IN OHMS 1/4 WATT ±5%.
2. ALL CAPACITANCE VALUES ARE IN MICROFARAD.
3. DOTTED LINES INDICATE SUGGESTED MODULAR BREAKDOWN.
4. Q2, Q4, Q5 & Q6 ARE 2N1121
LAST NUMBERS USED
C24, CR5, P1, Q6, R30, S1, T2
FM TUNER
A TYPICAL SCHEMATIC DIAGRAM PROBLEM
An engineer wants a breakdown of the circuit shown at left to facilitate module assembly; each technician is to work on a separate module. Prepare four separate schematic parts of the FM tuner below and on the next three pages in the following order:
Sheet 1 Complete the started power supply below.
Sheet 2 Complete the started converter and 1st IF amplifier.
Sheet 3 Complete the started 2nd IF amplifier and limiter.
Sheet 4 Complete the started demodulator and audio.
Use your electronics template. For symbols use Appendix A. Do not show the dotted lines.
The schematic part designations at left is drawn in accordance with the ASA manner [American Standards Association].
COMPLETE THE STARTED POWER SUPPLY OF THE FM TUNER BELOW.
P1
S1
T2
117 V
60 CPS
0V
+C24
200
15V
-12V
TITLE SCHEMATIC DIAGRAM - FM TUNER, POWER SUPPLY
DWG. NO. SCH-3
NAME
DATE
COURSE
GRADE
SCALE ~
SHEET 1 OF 4
PAGE 27

Exercise. Complete the started **converter** and **1st IF amplifier** below. Use sheet 1 (page 27) for reference.

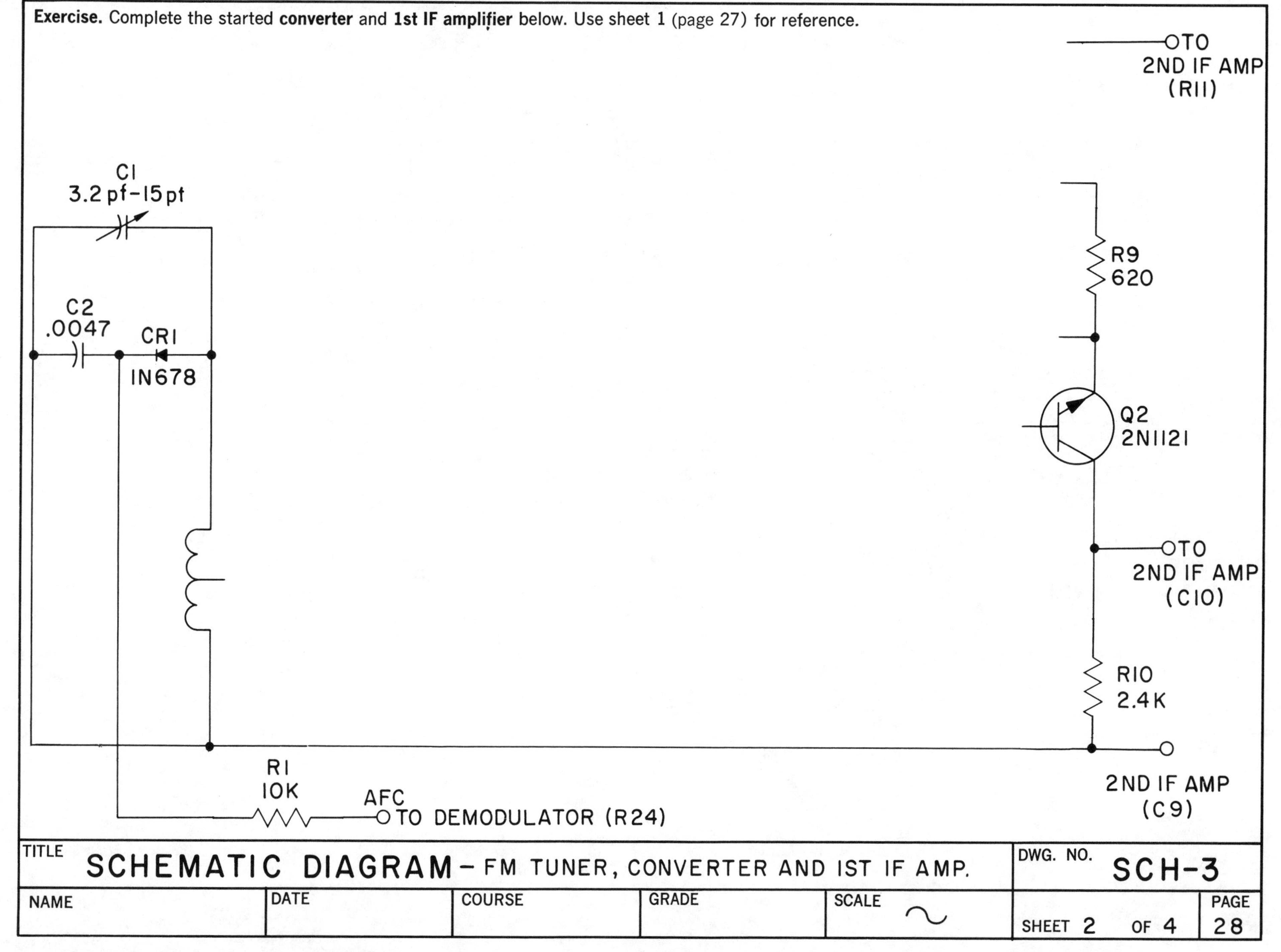

Exercise. Complete the started **2nd IF amplifier** and **limiter** below. Use sheet **1** (page 27) for reference.

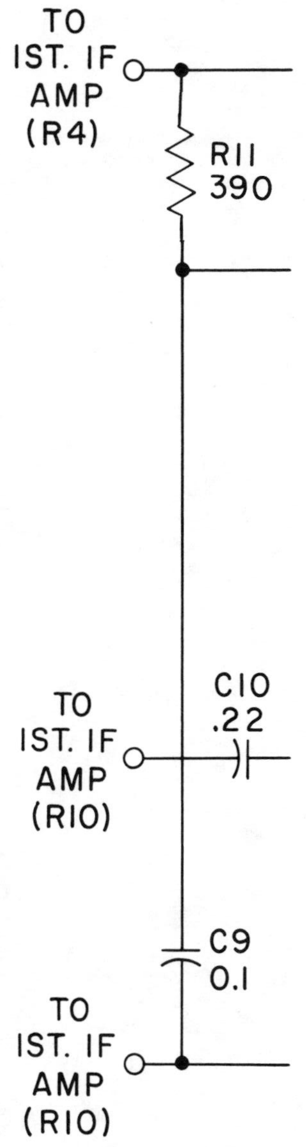

TITLE SCHEMATIC DIAGRAM – FM TUNER, 2ND IF AMP. AND LIMITER					DWG. NO. SCH-3	
NAME	DATE	COURSE	GRADE	SCALE ~	SHEET 3 OF 4	PAGE 29

Exercise. Complete the started **demodulator** and **audio** below. Use sheet 1 (page 27) for reference.

TO LIMITER (R19)

R21 3.3K

TITLE SCHEMATIC DIAGRAM—FM TUNER, DEMODULATOR AND AUDIO					DWG. NO. SCH-3	
NAME	DATE	COURSE	GRADE	SCALE	SHEET 4 OF 4	PAGE 30

CABLE DRAWINGS

The typical cable drawing at right has all the necessary information to identify parts and to assemble and properly install the cable.

In general, code numbers are brief notations for full descriptions. For example,

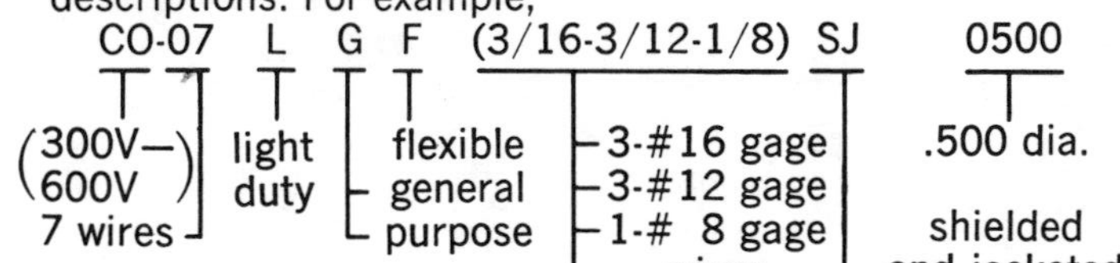

MS3101E-24-16P = connector-plug (male) which requires no clamp, bushing, or insert numbers.
MS25036-3 or MS25036-8 = lug types
MW-C12(65)J = wire type (J = Jacketed)
└Medium Wall-Copper #12 gage (.0808D)65 strand
TB2/3-J16/1-6 = wire I.D. and destination
└Jacketed #16(0508D) brown-blue (1-6:see Color Code Appendix E)
└Terminal Board #2/terminal #3
V2/5-J8/6 = wire I.D. and destination
└Jacketed #8 (.1285) blue = plate
└tube #2/terminal #5 = plate (anode)

Exercise 1. From code #HW-C8 (133)J, HW = heavy wall; C = ______________________

8 = ______________ J = ______________

Exercise 2. From the schematic, 3 BR-BL/16 means wire #3 is colored __________ and is/__________ (Gage ?)

Exercise 3. Wire #TB2/4-J16/2-3 is colored and __________. Its destination is __________

(Hint: See wire table)

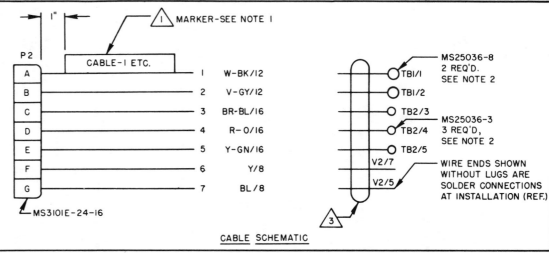

1" P2 A B C D E F G
MS3101E-24-16
CABLE-1 ETC.
△1 MARKER-SEE NOTE 1
1 W-BK/12
2 V-GY/12
3 BR-BL/16
4 R-O/16
5 Y-GN/16
6 Y/8
7 BL/8
TB1/1
TB1/2
TB2/3
TB2/4
TB2/5
V2/7
V2/5
MS25036-8 2 REQ'D. SEE NOTE 2
MS25036-3 3 REQ'D, SEE NOTE 2
WIRE ENDS SHOWN WITHOUT LUGS ARE SOLDER CONNECTIONS AT INSTALLATION (REF.)
△3

CABLE SCHEMATIC

WIRE TABLE

NO.	TYPE	LENGTH	FROM	TO	STRIP	REMARKS
1	MW-C12(65)J	6'-4½"	P2-A	TB1/1-J12/9-0	¼-½	ALL WIRES PER SPEC MIL-W-16878 TYPE B
2	MW-C12(65)J	6'-4½"	P2-B	TB1/2-J12/7-8	¼-½	
3	MW-C16(26)J	6'-5"	P2-C	TB2/3-J16/1-6	¼-⅜	
4	MW-C16(26)J	6'-5"	P2-D	TB2/4-J16/2-3	¼-⅜	
5	MW-C16(26)J	6'-5"	P2-E	TB2/5-J16/4-5	¼-⅜	
6	HW-C8(133)J	6'-5½"	P2-F	V2/7-J8/4	¼-⅜	
7	HW-C8(133)J	6'-5½"	P2-G	V2/5-J8/6	¼-⅜	

△3 CABLE TUBING: ALPHA #PVC-105-1/2" BY 6 FT. LONG PER MIL-1-631C OR EQUIVALENT

2. LUGS FURNISHED UNATTACHED, TO BE CRIMPED AT INSTALLATION.

△1 INK STAMP 1/8 CHARACTERS WITH BLACK INK AS FOLLOWS: CABLE-1 CO-07LGF (3/16 - 2/12 - 2/8) SJ 0500 ON CABLE TUBING AND COAT WITH CLEAR PLASTIC SPRAY 1" FROM PLUG AS SHOWN.

NOTES:

TITLE CABLE DRAWING—CABLE SCHEMATIC AND WIRE TABLE					DWG. NO. CD-1	
NAME	DATE	COURSE	GRADE	SCALE	SHEET 1 OF 3	PAGE 31

The same cable as was used on sheet 1 is shown here **orthographically** and **schematically.** This is an **assembly drawing** which requires a parts list to obtain parts and a schematic (or wire table) to properly assemble and install the cable. The section view suggests how the tubing size is selected. Notice here the use of the **arrow** rather than **block** symbol for **schematic multiplug.** (These are interchangeable).

Exercise 1. From sheet 1 complete schematic at right. Complete notes △1 and 2. Have note 3 explain which wires are to be stripped and by how much. Note 4 should mention the wire specifications.

Exercise 2. From code number V2/7-J8/4

V2 =
7 =
J =
8 =
4 = YELLOW; CATHODE

Exercise 3. From code number CO-22HOE(10/8-12/10)SJ1500

E = extra flexible
O = oil resistant
H =
22 =
12/10 =
1500 =

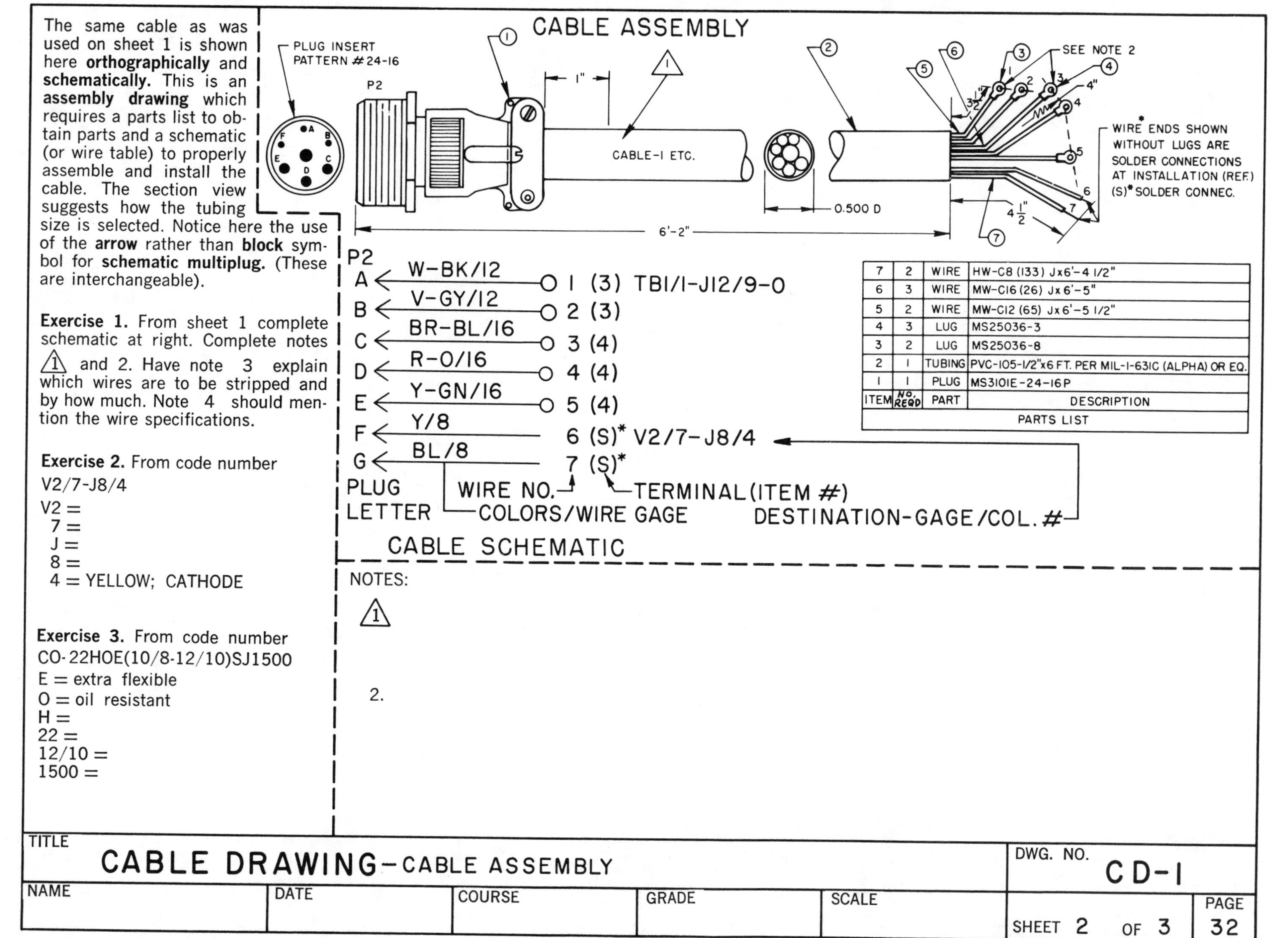

7	2	WIRE	HW-C8 (133) Jx6'-4 1/2"
6	3	WIRE	MW-C16 (26) Jx6'-5"
5	2	WIRE	MW-C12 (65) Jx6'-5 1/2"
4	3	LUG	MS25036-3
3	2	LUG	MS25036-8
2	1	TUBING	PVC-105-1/2"x6 FT. PER MIL-I-631C (ALPHA) OR EQ.
1	1	PLUG	MS3101E-24-16P
ITEM	NO. REQD	PART	DESCRIPTION
PARTS LIST			

NOTES:

△1

2.

TITLE **CABLE DRAWING**–CABLE ASSEMBLY					DWG. NO. CD-1	
NAME	DATE	COURSE	GRADE	SCALE	SHEET 2 OF 3	PAGE 32

Make a **cable drawing** similar to that on sheet 2 with same plug "shell" (MS3101E) and insert #24-27P, which carries seven #16 wires (in the same pattern as #24-16P). Terminate all wires with the MS25036-3 lug 4″ beyond tubing. Tubing and plug length to be 8′-9″. All wires #1 through #7 to go from P2 to TB3, terminals #1 through #7 (wire #1 to terminal #1, wire #2 to terminal #2, etc.).

Have all wires **jacketed with color numbers** to match wire and terminal numbers. Complete "block" multiplug schematic and parts list. Use basically the same style format and notes as on sheet 2 (but call out revised **length of wires** and **tubing** plus revised **wire colors** and **size).**

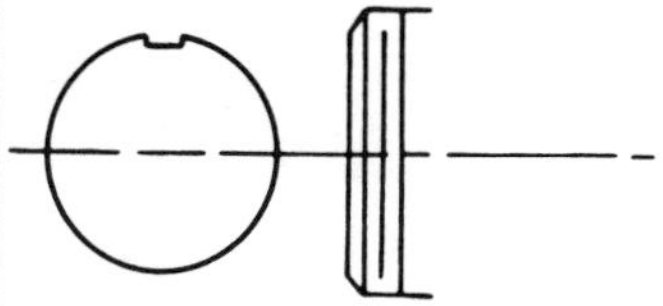

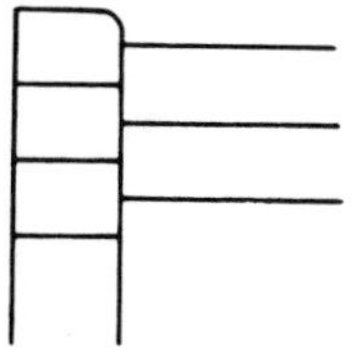

PARTS LIST			

2.

1

NOTES:

TITLE CABLE DRAWING					DWG. NO. CD-1	
NAME	DATE	COURSE	GRADE	SCALE	SHEET 3 OF 3	PAGE 33

MILITARY STANDARDS (MS)

Military Standards become known to draftsmen by many code numbers. Examples are:

ANSI Y32.2 (electrical and electronics symbols)

MIL-A-8625 (anodize)

QPL-641-16 (jacks, telephone)

JAN-S-28 (sockets, electronic tube)

QQ-A-327 (aluminum sheet and plate)

MS35221-45 (screw, pan head)

In addition, there are hundreds whose code letters start with AN-, ANA-, EI-, AND-, and many others. These various Military Standards permit the electromechanical draftsman to use a **short code number** to represent a lengthy description or detailed drawing.

In industry the electronics draftsman spends a great deal of his time looking up electronic components, hardware and specs. Basically, the MS (Military Standard) number describes one of the following:

A booklet of general information

An engineering material

A fabricating process

An electronic component

A piece of hardware

Handling procedure (assembly, installation, or operation of equipment)

See examples of typical pages of Military Standards in Appendix C, pages 124-127.

Exercises. In order to answer the exercises below, refer to Appendix C, pages 124–127 (Military Standards).

1. Complete MS35221-14 callout below with its material and finish.

   #4-40NC X 5/16 LONG, PAN HEAD SCREW

   Material ________ Finish ________

2. What is MS35649-84?

   ____

3. What is the outside diameter of MS35337-3?

   ____

4. What is the MS number of the following:

   #2, WASHER, FLAT GEN. PURPOSE CRES ________

   #2, WASHER, FLAT GEN. PURPOSE BRASS ________

   #6, LOCK WASHER, SPLIT, LIGHT SERIES, PLAIN ________

5. What information is given in the first four columns of MS15795?

   ____

   What are the rest of the columns used for?

   ____

6. What are the MS numbers of all the hardware shown in Fig. 1?

   (HARDWARE MATERIAL — CRES)

FIG. 1

SCREW ________

NUT ________

WASHER, FLAT ________

WASHER, LOCK ________

TITLE MILITARY STANDARDS—INTRODUCTION					DWG. NO. MS-1	
NAME	DATE	COURSE	GRADE	SCALE NONE	SHEET 1 OF 2	PAGE 34

A typical callout of a fastening using MS numbers might read as follows;

Example

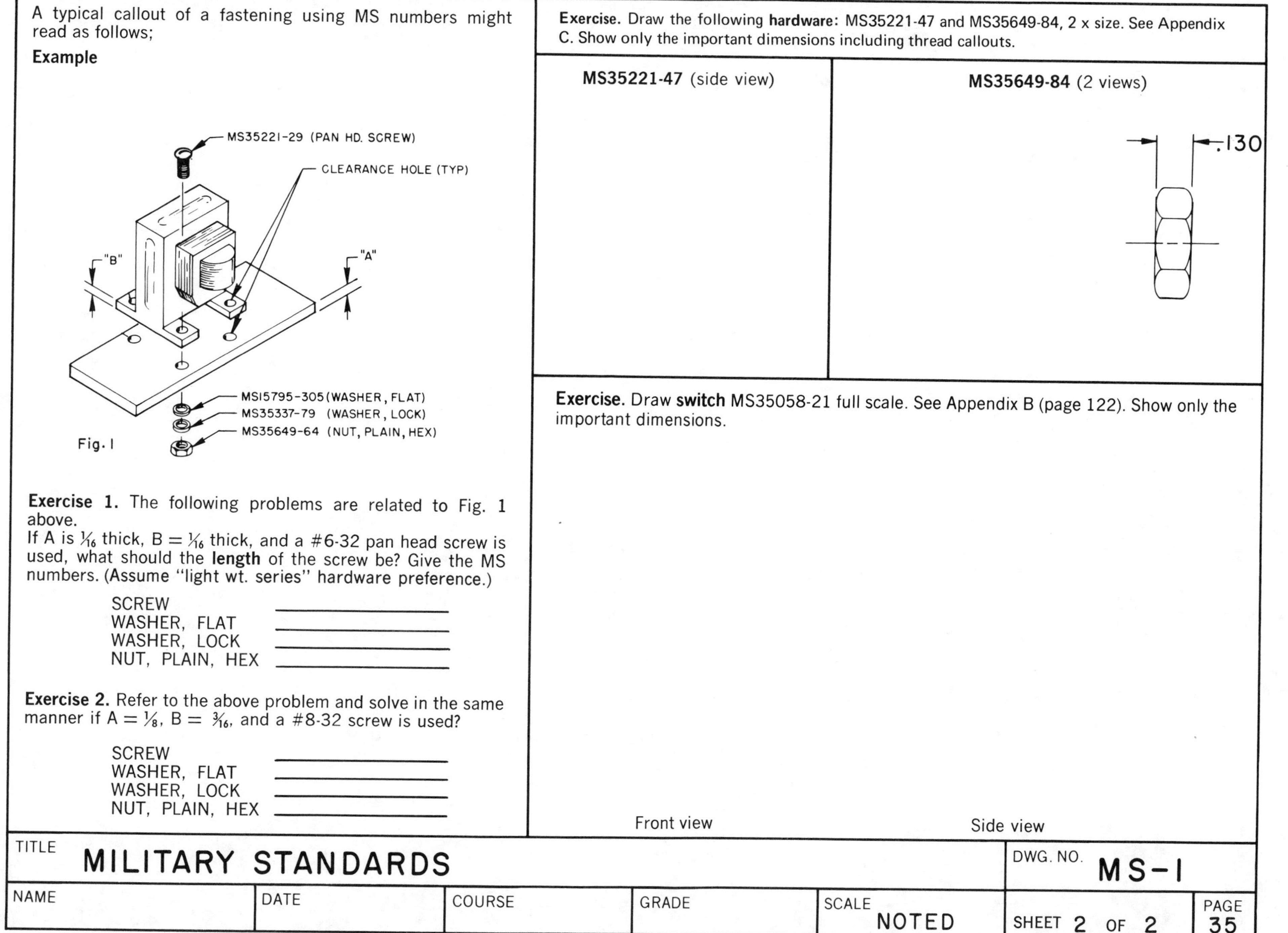

Fig. 1

Exercise 1. The following problems are related to Fig. 1 above.
If A is $\frac{1}{16}$ thick, B = $\frac{1}{16}$ thick, and a #6-32 pan head screw is used, what should the **length** of the screw be? Give the MS numbers. (Assume "light wt. series" hardware preference.)

SCREW ________
WASHER, FLAT ________
WASHER, LOCK ________
NUT, PLAIN, HEX ________

Exercise 2. Refer to the above problem and solve in the same manner if A = $\frac{1}{8}$, B = $\frac{3}{16}$, and a #8-32 screw is used?

SCREW ________
WASHER, FLAT ________
WASHER, LOCK ________
NUT, PLAIN, HEX ________

Exercise. Draw the following **hardware:** MS35221-47 and MS35649-84, 2 x size. See Appendix C. Show only the important dimensions including thread callouts.

MS35221-47 (side view)

MS35649-84 (2 views)

Exercise. Draw **switch** MS35058-21 full scale. See Appendix B (page 122). Show only the important dimensions.

Front view

Side view

TITLE MILITARY STANDARDS					DWG. NO. MS-1	
NAME	DATE	COURSE	GRADE	SCALE NOTED	SHEET 2 OF 2	PAGE 35

PRINTED CIRCUIT PATTERN

The length of the conductors between various lands shall be held to a minimum.

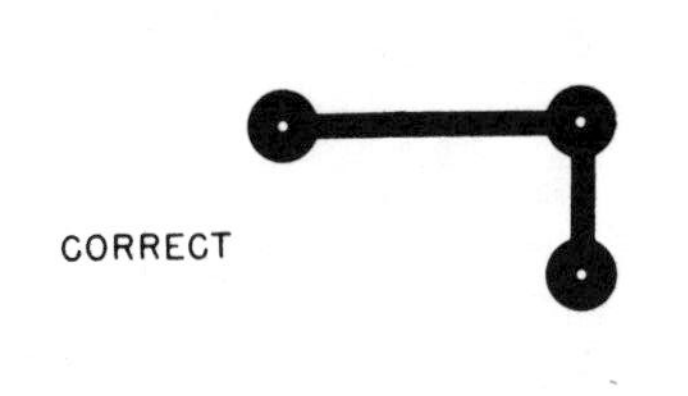

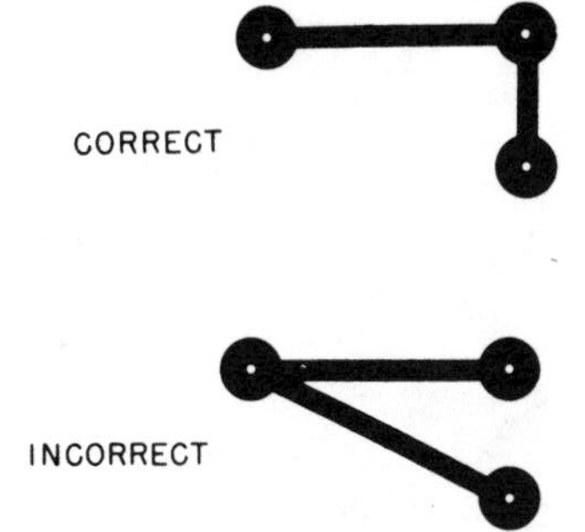

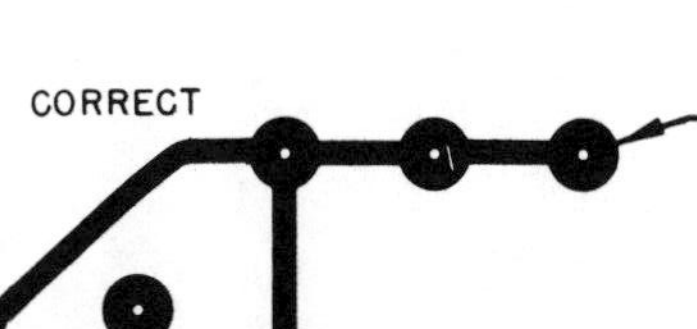

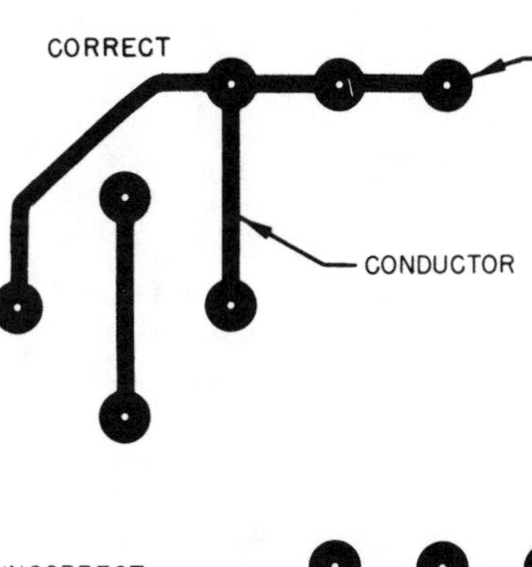

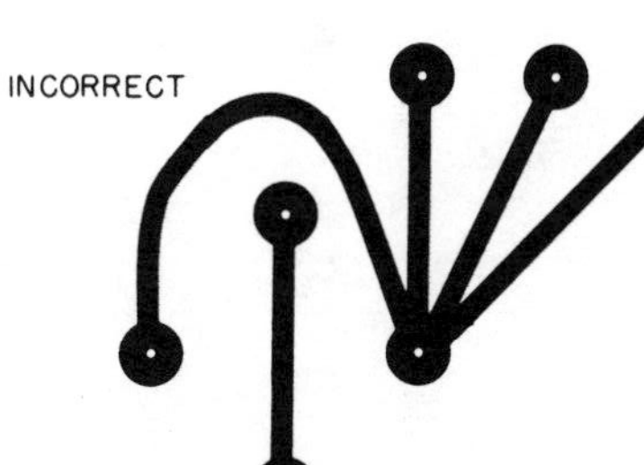

Exercise 1: Connect the lands given in Figs. 1, 2, and 3 with a 1/16″ wide black pressure-sensitive tape. Leave 1/16 minimum spacing from edge of board and component holes. If no tape is available, use pencil and draw the conductors 1/16 wide (A tee, see Fig. 4, with fillets is optional.)

Fig. 1

Fig. 2

Fig. 3

Exercise 2. With a 1/16″ wide tape connect the following in Fig. 4.

- All lands marked A join together.
- All lands marked B join together.
- Connect nearest A hole to remaining connector tab (as shown with B hole to tab).

Make all connections on one side of board only, as shown, and keep the length of the conductors to a minimum.

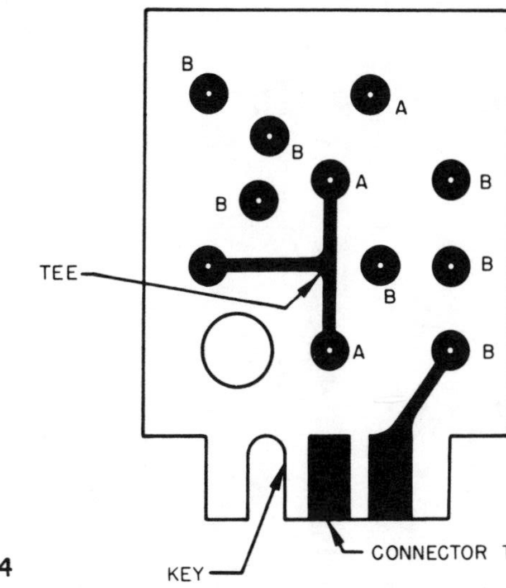

Fig. 4

Exercise 3. With a 1/16″ wide tape connect the following in Fig. 5:

- All 1 lands together to **Grd** (ground).
- All 2 lands together to **Out.**
- All 6 lands together.

Connect transistor as follows:

- Base B to No. 4 tab.
- Collector C to No. 5 tab.
- Emitter E to land 3.

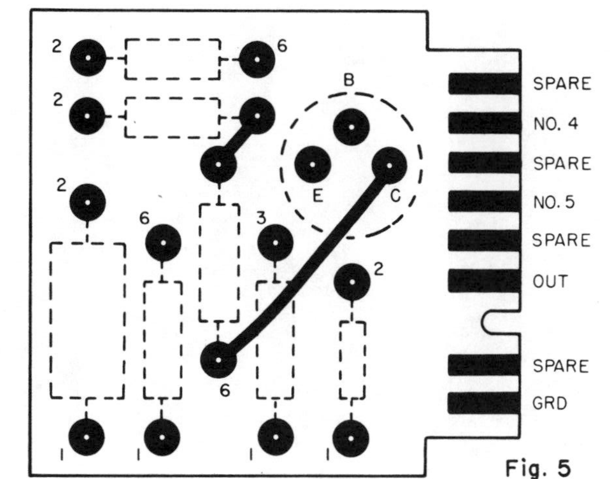

Fig. 5

TITLE PRINTED CIRCUIT BOARD – P.C. PATTERN (SINGLE SIDED)					DWG. NO. PCB-1A	
NAME	DATE	COURSE	GRADE	SCALE 2 X SIZE	SHEET 1 OF 1	PAGE 36

TRANSISTOR AND DIODES IN PRINTED CIRCUIT BOARD (PCB)

Special attention should be given to laying out transistors and diodes in printed circuit boards. **Transistor leads** look different when viewed from the bottom (view *a*) than when viewed from the top (view *b*), as shown in Fig. 1.

BOTTOM VIEW

EMITTER (E) BASE (B) TAB COLLECTOR (C)

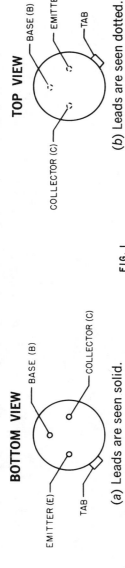

(*a*) Leads are seen solid.

TOP VIEW

COLLECTOR (C) BASE (B) EMITTER (E) TAB

(*b*) Leads are seen dotted.

FIG. I

The secret lies in the **tab** of the transistor. The emitter (E) is closest to the tab of the transistor. The **diode** is easier to lay out. Always observe in the schematic to which side the cathode is pointed. In the layout the diode should be installed in its proper orientation or
It is important that the polarity of the diode be shown.

Exercise: From the schematic shown below make a layout of the components as seen from the top of the board (the circuit will be dotted) and as seen from the circuit side (the components will be seen dotted). Show all missing components and lines. Show **transistor tabs** in bottom view. Lay out one component lead per land.
Component outlines will be found in Appendix B.

Remember: Always count counter-clockwise from tab: "emitter, base, collector" to identify transistor elements in top view.

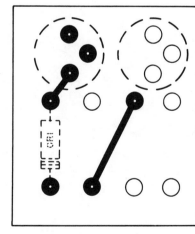

SCHEMATIC DIAGRAM

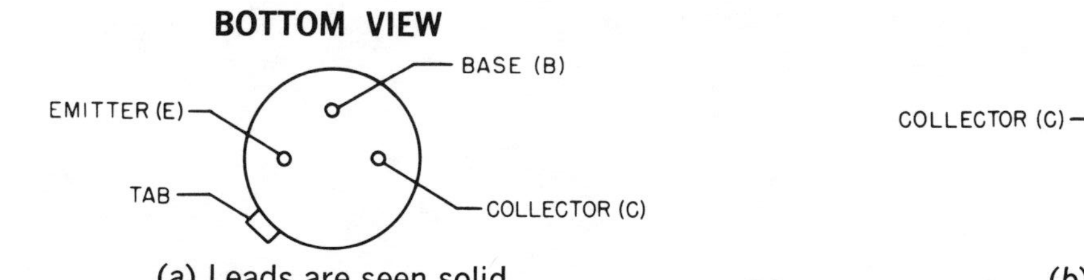

TOP VIEW
(component side)

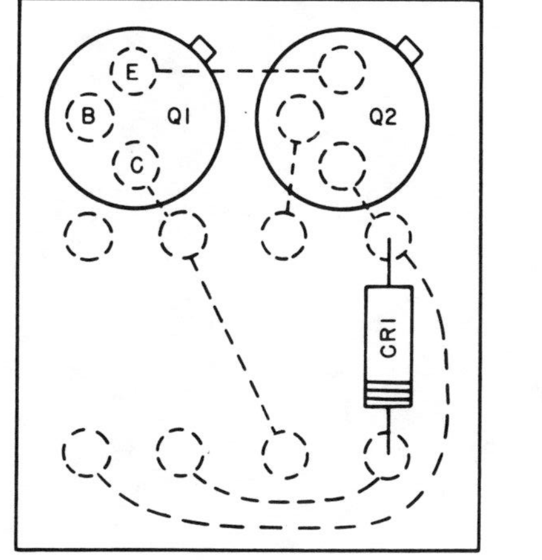

SIDE VIEW

BOTTOM VIEW
(circuit side)

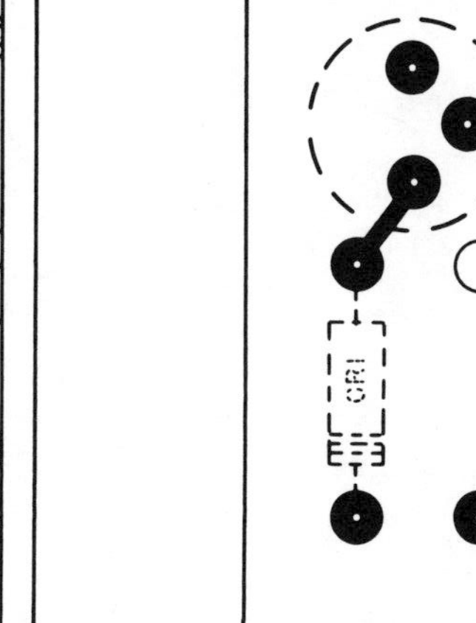

TITLE PRINTED CIRCUIT BOARD–TRANSISTORS AND DIODES					DWG. NO. PCB-1B	
NAME	DATE	COURSE	GRADE	SCALE 2 X SIZE	SHEET 1 OF 1	PAGE 37

A TYPICAL, SINGLE-SIDED PCB (Printed circuit board)

STEP-BY-STEP FABRICATION

Printed circuit boards are fabricated in five principal steps, requiring five types of drawings:

1. **Schematic** 2. **Layout** 3. **Master** (Artwork) 4. **Drill & Contour** and 5. **Assembly** with **List of Materials.**

From specifications on a Schematic as shown at the upper right, a Layout started at the lower right can be drawn using components outlines similar to Appendix B.

Given: Schematic at right with component outline and Appendix B.

Problem: Prepare the drawings required to fabricate a PCB. Mount all components on a single-sided board 1.0 × 1.0 × .062 thick, and provide three PAN HEAD (#2-56 thread) SCREWS for installation. Assume ℄ - ℄ Tol. = .005, and clearance hole on clearance hole type mounting.

Solution: Steps 1 and 2 are described at right while steps 3 through 5 are developed on page 39 & 40.

HOW TO PREPARE THE LAYOUT

Lay out your board and components double size for photographing. Remember that the transistor is viewed from top (see previous exercise). *In dotted line or red pencil* show the conductors between lands. Show the screw head, three places. Mark components and polarity of diodes as shown. *Always make a neat, accurate layout.*

The layout should show screws (as shown in PCB-2) or any other fastener which will occupy space on the P.C. Board and displace circuitry or component area. Although the screws have to be shown in the layout, they are not considered part of the assembly and should not appear in the List Of Materials.

SCHEMATIC OF PCB-2

IN, CR1 IN617, CR2 IN617, C1 50, Q1 2N1304, OUT, R1 470K, TERM. 1, R2 6.8K, B+

NOTE:
RESISTORS = 1/4 W, 5%
CAPACITOR = mmf, 500V

Step 1: SCHEMATIC (PCB-2)
Always draw the schematic first. Since this is a Sample Exercise, the schematic is shown completed.

Step 2: LAYOUT EXERCISE (PCB-2)
Complete the Layout.

1. How many electrical components are there?__________

LAYOUT OF PCB-2

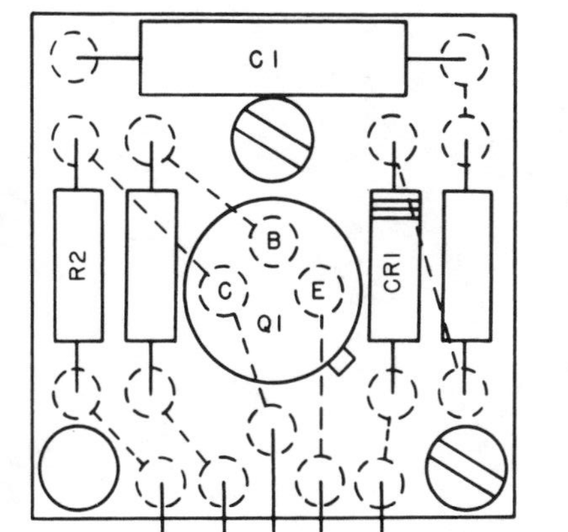

2. Two components are not identified. Show them in the layout. (See schematic diagram.)
3. Show polarity of all diodes.
4. One screw head is not shown finished. Complete it.
5. Identify all leads coming out of the board.
6. Check your layout against the schematic and fill in the missing dotted line.

TITLE PRINTED CIRCUIT BOARD–SCHEMATIC AND LAYOUT (SINGLE SIDED)					DWG. NO. PCB-2	
NAME	DATE	COURSE	GRADE	SCALE 2 X SIZE	SHEET 1 OF 3	PAGE 38

Step 3: PCB MASTER (Artwork)

Generally, the layout made in Step 2 is turned over and a sheet of mylar is placed on the reverse side. The layout is usually made on semitransparent vellum whose PCB pattern is visible from the back side through the mylar. The PCB pattern is now duplicated on the mylar with black tape conductors and/or decal lands. If you had made a master of PCB-2, it would be similar to the sample shown below.

The master (artwork) should show the following:

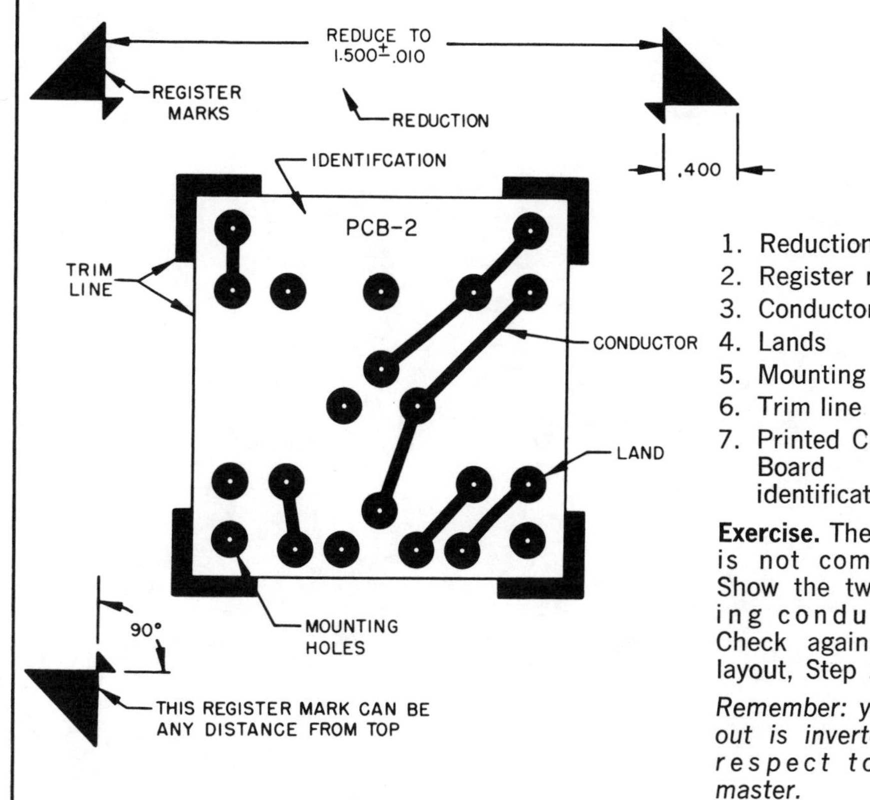

1. Reduction
2. Register marks
3. Conductors
4. Lands
5. Mounting holes
6. Trim line
7. Printed Circuit Board identification

Exercise. The sample is not completed. Show the two missing conductors. Check against your layout, Step 2.

Remember: your layout is inverted with respect to your master.

Step 4: PRINTED CIRCUIT BOARD DRILLED (PCB-2)

This step is needed so that the printed circuit board can be fabricated after the etching is done. Include the following data in all exercises to come:

1. Show all hole-diameter callouts.
2. Show board mounting hole center dimensions.
3. Show the shape and size of finished board dimensions.
4. **Material:** .062 thick Epoxy glass cloth laminate with .0027 CU one side (CU = copper)
5. **Finish:** Gold flash on etched lettering and conductors.

Exercise: See illustration below.

1. What should the .XX dimension be? ____________
2. What are the A holes for? ____________
   How many A holes are there? ____________
3. What are the B holes for? ____________
   How many B holes are there? ____________
4. What equations (see p. 128, ex. 3) were used to find .096 DIA "B" holes? ____________
5. Complete the missing dimensions.
6. Show the two missing conductors as in step 3.

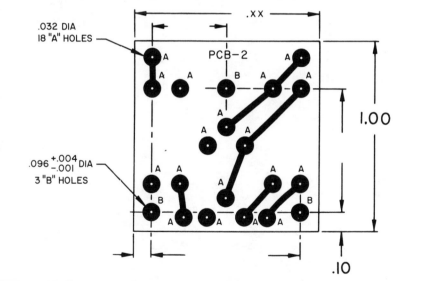

.XX
.032 DIA 18 "A" HOLES
PCB–2
$.096^{+.004}_{-.001}$ DIA 3 "B" HOLES
1.00
.10

TITLE PRINTED CIRCUIT BOARD–MASTER AND DRILLED (SINGLE SIDED)					DWG. NO. PCB-2	
NAME	DATE	COURSE	GRADE	SCALE 2 X SIZE	SHEET 2 OF 3	PAGE 39

Step 5: PRINTED CIRCUIT BOARD ASSEMBLY (PCB-2)

The following items must be shown:

1. Assembly of all components as seen mounted on board. It will be the same as your layout, except that the dotted line and lands will not be seen since it is on the opposite side of the board.

2. All components and identification. *Always make it clear so that anyone could assemble the board from your assembly drawings.*

3. The assembly drawing should always be accompanied by a complete list of material, as shown below.

ITEM	NO. REQ.D	REFERENCE DESIGNATION	DESCRIPTION	MANUFACTURER & PART NO. OR MIL TYPE DESIGNATION
7			RESISTOR	6.8K, ¼W, 5% RC07GF682J
6				
5				
4	2	CR1, CR2	DIODE	
3	1	C1	CAPACITOR	50 MMF, 500V, CM
2	1	PCB-2	PRINTED BOARD	1.00 x 1.00 x .062 EPOXY GLASS
1	X	———————	SCHEMATIC	SHEET 1 of 3

LIST OF MATERIAL

Exercise

1. Complete the assembly drawing below. Show all component identifications.

2. Complete the list of material (see Appendix B and schematic diagram on page 38).
   Resistors are listed on page 115 on page 115 (MIL STYLE RC07).

   Capacitors are listed on page 112 (MIL TYPE CM-15).

   Diodes are listed on page 114.

   Transistors are listed on page 118.

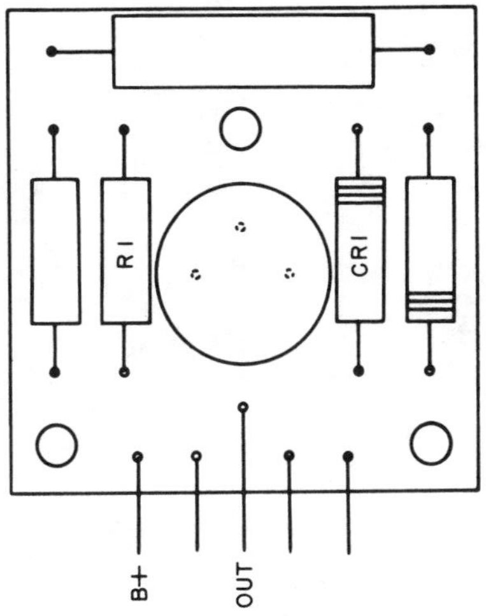

TITLE PRINTED CIRCUIT BOARD—ASS'Y AND LIST OF MATERIAL (SINGLE SIDED)					DWG. NO. PCB-2	
NAME	DATE	COURSE	GRADE	SCALE 2 X SIZE	SHEET 3 OF 3	PAGE 40

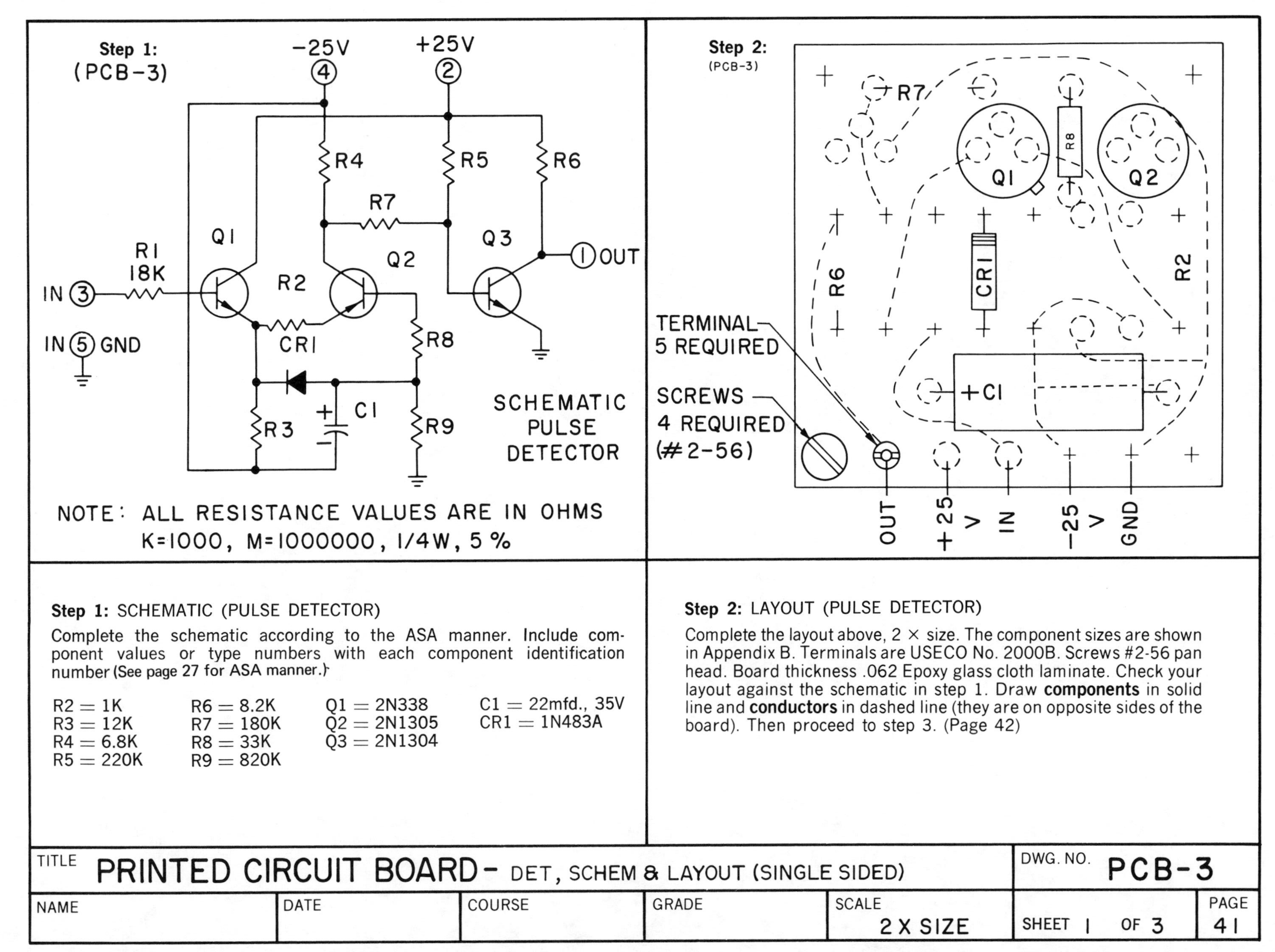

Step 1: SCHEMATIC (PULSE DETECTOR)

Complete the schematic according to the ASA manner. Include component values or type numbers with each component identification number (See page 27 for ASA manner.)

R2 = 1K	R6 = 8.2K	Q1 = 2N338	C1 = 22mfd., 35V
R3 = 12K	R7 = 180K	Q2 = 2N1305	CR1 = 1N483A
R4 = 6.8K	R8 = 33K	Q3 = 2N1304	
R5 = 220K	R9 = 820K		

Step 2: LAYOUT (PULSE DETECTOR)

Complete the layout above, 2 × size. The component sizes are shown in Appendix B. Terminals are USECO No. 2000B. Screws #2-56 pan head. Board thickness .062 Epoxy glass cloth laminate. Check your layout against the schematic in step 1. Draw **components** in solid line and **conductors** in dashed line (they are on opposite sides of the board). Then proceed to step 3. (Page 42)

TITLE PRINTED CIRCUIT BOARD - DET, SCHEM & LAYOUT (SINGLE SIDED)					DWG. NO. PCB-3	
NAME	DATE	COURSE	GRADE	SCALE 2 X SIZE	SHEET 1 OF 3	PAGE 41

Step 3: PRINTED CIRCUIT BOARD MASTER (ARTWORK)

Complete the master, following the same method used in sample exercise, PCB-2. Show reduction. (Reduce to 2.000 ± .010, which will reduce the board to half size giving a 1.75 × 1.75 full-size board.) Show register marks, conductors, etc. (**Hint:** Show three register marks, four trim lines and one reduction note. Omit nomenclature callouts of page 39.)

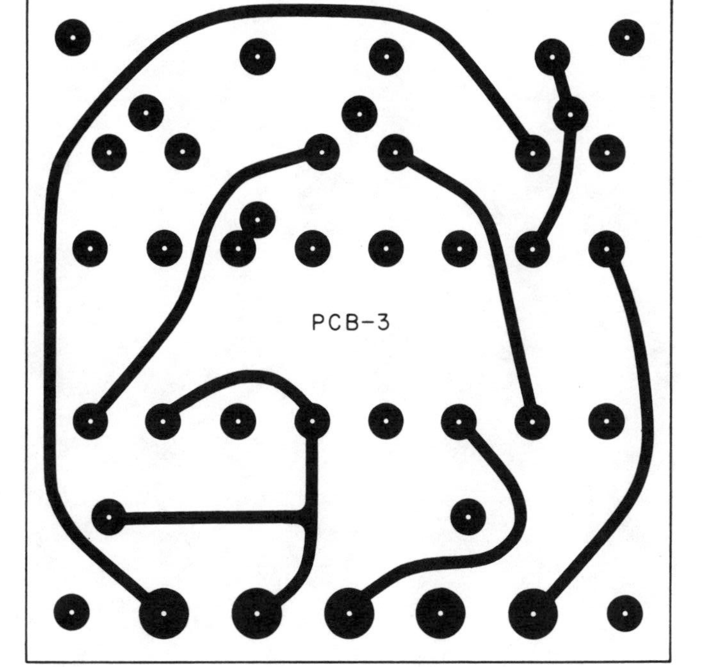

Step 4: PRINTED CIRCUIT BOARD DRILLED

Follow the same method used in sample exercise PCB-2. Complete the drawing of the board below. For clarity do not show conductors again but show all dimensions necessary for fabrication of board.

Hole diameter

A (land hole) = .032 Dia.

C (terminal mounting holes) = .065 $^{+.003}_{-.001}$ Dia.

Calculate **clearance B holes**, four places for #2-56 pan head screws (Use Formula, Appendix D — use "B" ℄ - ℄ tol. = ±.005 — tapped hole mtd).

B B PCB-3 B B

TITLE PRINTED CIRCUIT BOARD—DETECTOR, MASTER AND DRILLED BOARD					DWG. NO. PCB-3	
NAME	DATE	COURSE	GRADE	SCALE 2 X SIZE	SHEET 2 OF 3	PAGE 42

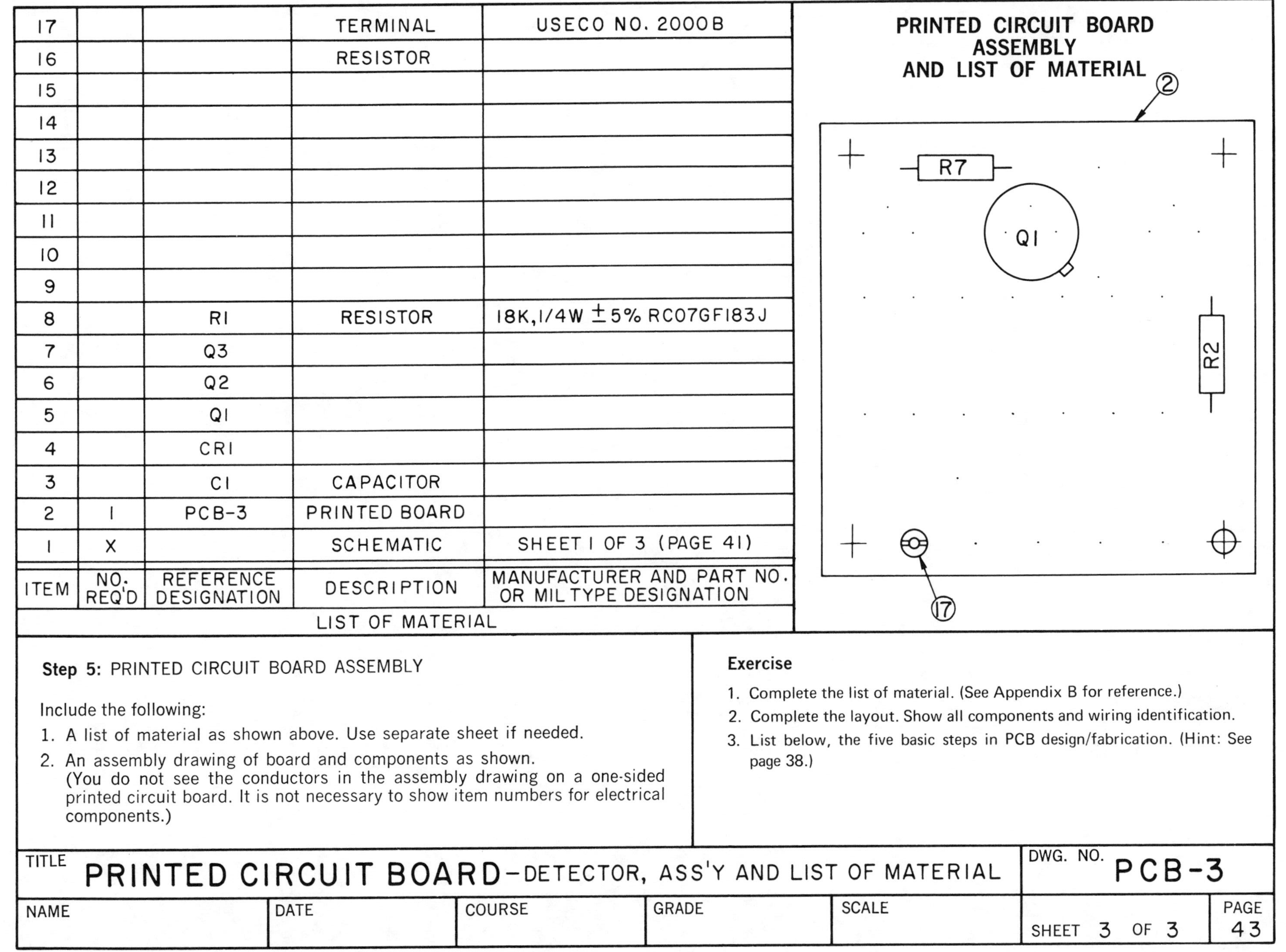

ITEM	NO. REQ'D	REFERENCE DESIGNATION	DESCRIPTION	MANUFACTURER AND PART NO. OR MIL TYPE DESIGNATION
17			TERMINAL	USECO NO. 2000B
16			RESISTOR	
15				
14				
13				
12				
11				
10				
9				
8		R1	RESISTOR	18K, 1/4W ±5% RC07GF183J
7		Q3		
6		Q2		
5		Q1		
4		CR1		
3		C1	CAPACITOR	
2	1	PCB-3	PRINTED BOARD	
1	X		SCHEMATIC	SHEET 1 OF 3 (PAGE 41)

LIST OF MATERIAL

Step 5: PRINTED CIRCUIT BOARD ASSEMBLY

Include the following:

1. A list of material as shown above. Use separate sheet if needed.
2. An assembly drawing of board and components as shown.
   (You do not see the conductors in the assembly drawing on a one-sided printed circuit board. It is not necessary to show item numbers for electrical components.)

Exercise

1. Complete the list of material. (See Appendix B for reference.)
2. Complete the layout. Show all components and wiring identification.
3. List below, the five basic steps in PCB design/fabrication. (Hint: See page 38.)

TITLE PRINTED CIRCUIT BOARD–DETECTOR, ASS'Y AND LIST OF MATERIAL					DWG. NO. PCB-3	
NAME	DATE	COURSE	GRADE	SCALE	SHEET 3 OF 3	PAGE 43

DOUBLE-SIDED PRINTED CIRCUIT BOARDS

Double-sided PCBs have the following features:

1. They are more expensive to manufacture than single-sided PCBs.
2. They are used where space is limited.
3. In double-sided boards the components are on *one or both* sides of the board.
4. Conductors are on *both sides* of the board.
5. An accurate layout is required in order to line up the front and back faces of the board; therefore, tool holes are required for machining.

CONNECTING THE FRONT AND BACK FACE OF THE BOARD

The four general methods for accomplishing through connection between two sides of a double-sided PCB are (1) leads, (2) eyelets, (3) plated-through holes, and (4) terminals.

Exercise (Step 1 and Step 2 PCB-4)
From the same schematic diagram Pulse Detector and list of material in Exercise PCB-3, Fabricate a double-sided PCB.

Given:

1. **Schematic,** Exercise PCB-3, Step 1. (Page 41)
2. **List of material,** Exercise PCB-3, Step 5. (Page 43)
3. **PCB-4,** 1.750 x 1.250 x .062 thick. (Notice the board is now smaller in size than PCB-3.)

PROBLEM: Complete the started layout of PCB-4 below. Show transistor tabs in proper location. **Three components,** shown with hidden lines, are mounted on the back side of the board.

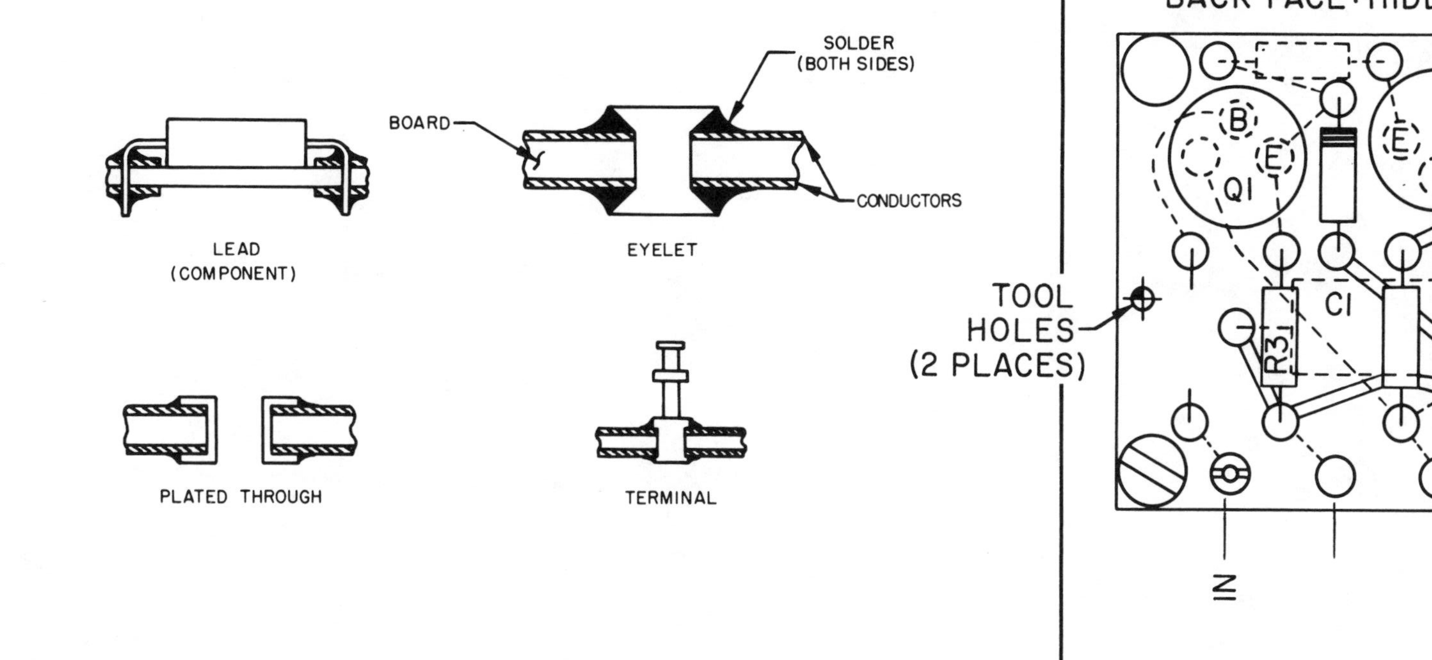

TITLE PRINTED CIRCUIT BOARD – DOUBLE SIDED, BOARD LAYOUT					DWG. NO. PCB-4	
NAME	DATE	COURSE	GRADE	SCALE 2 X SIZE	SHEET 1 OF 4	PAGE 44

Step 3: MASTER (ARTWORK) PCB-4

The method is the same as in the single-sided board, except that in the double-sided PCB **two masters** (two pieces of artwork) are needed for photographing.

Exercise: PCB-4

Complete the two masters (both sides of PCB-4) below. Follow the same method used in preparing the master (artwork) of PCB-3 (step 3).

The **front-face master** is prepared from the layout with *solid lines.*

The **back-face master** is prepared from the layout with *hidden lines.*

Show the correct reduction of PCB-4 for both sides (**Hint:** board size is 1.750 x 1.250 x .062) Tool holes are already shown in both masters.

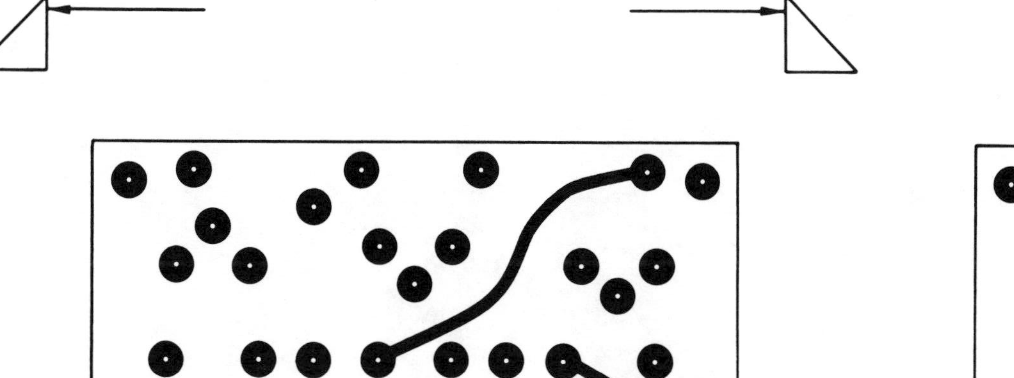

FRONT FACE

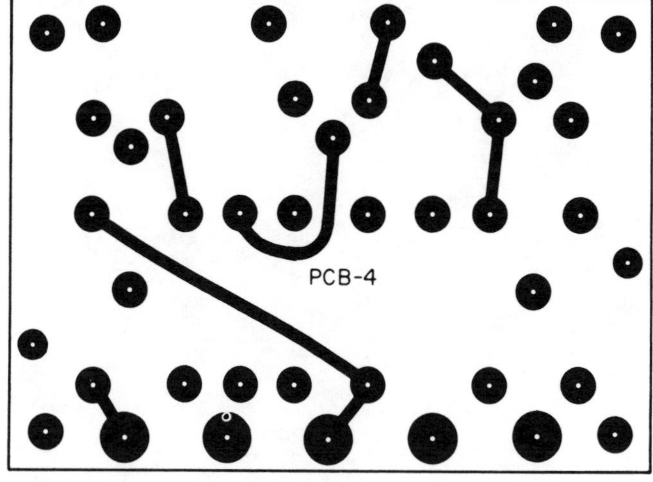

BACK FACE

TITLE PRINTED CIRCUIT BOARD—DOUBLE SIDED, BOARD MASTERS					DWG. NO. PCB-4	
NAME	DATE	COURSE	GRADE	SCALE 2 X SIZE	SHEET 2 OF 4	PAGE 45

Step 4: PCB DRILLED (DOUBLE-SIDED PCB-4)

The method is the same as in a single-sided board. Choose the master with the identification PCB-4. Since the lands on both sides of the PCB are aligned, drilling may be done through either side.

Exercise: PCB-4

Complete the PCB-4 drilled drawing below in the same manner as was used in Exercise PCB-3 (step 4). Add the tool holes (.062 Dia.) and complete all the notes, hole chart, etc. Use hole chart instead of leader callouts.

(Hint: See page 39 for Material & Finish)

NOTES

MATERIAL : .062 THICK EPOXY____________

____

____

FINISH : GOLD___________________

__________________BOTH SIDES.

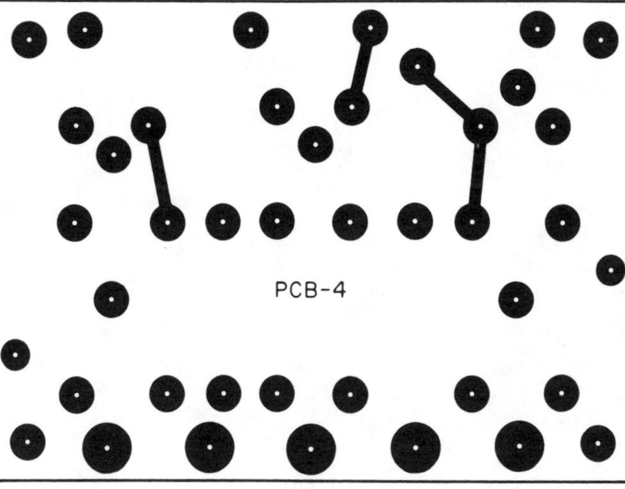

HOLE CHART			
DESCRIPTION	LETTER	DIA.	NO. REQ'D
LAND HOLES	A		
BOARD MTG. HOLE	B		
TERMINAL HOLE	C	$.065^{+.003}_{-.001}$	
TOOL HOLES	D		2

TITLE PRINTED CIRCUIT BOARD - DOUBLE SIDED, BOARD DRILLED					DWG. NO. PCB-4	
NAME	DATE	COURSE	GRADE	SCALE 2 X SIZE	SHEET 3 OF 4	PAGE 46

Step 5: ASSEMBLY (DOUBLE-SIDED PCB-4)

The assembly is prepared in the same manner as the single-sided board PCB-3 (step 5). The only difference is that in the double-sided PCB *two sides* are to be shown — the front and the back — with all the component identifications. Since this exercise has the same list of material as in PCB-3, it is not necessary to show it again.

Exercise: PCB-4

Complete the assembly drawing below by showing all the components. Some of the conductors have been omitted for clarity. Do not draw them in.

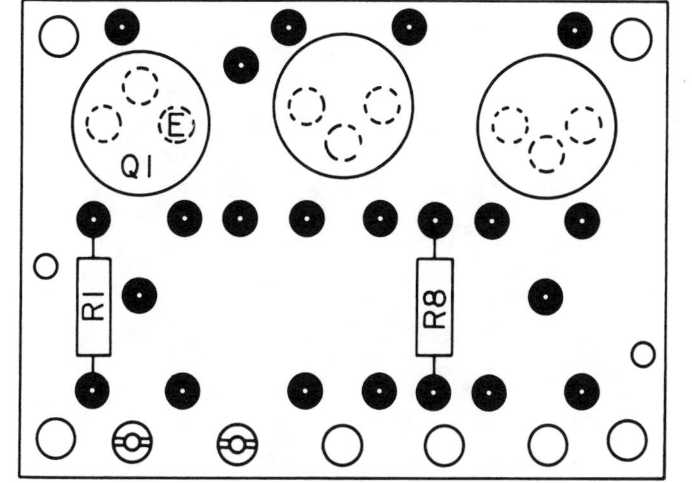

FRONT FACE

CI
+
−

BACK FACE

TITLE PRINTED CIRCUIT BOARD–DOUBLE SIDED, BOARD ASSEMBLY					DWG. NO. PCB-4	
NAME	DATE	COURSE	GRADE	SCALE 2X·SIZE	SHEET 4 OF 4	PAGE 47

Exercise: PCB-5 (PRINTED CIRCUIT BOARD)

From the schematic of the **Flip-Flop (PCB-5)** shown below, prepare all the drawings required (four more) to fabricate a **Single-Sided PCB-5** (see exercises PCB-3).

(**Note:** Student has the option of doing this page 48 problem or the easier page 51 problem first.)

SCHEMATIC DIAGRAM–PCB-5
FLIP–FLOP

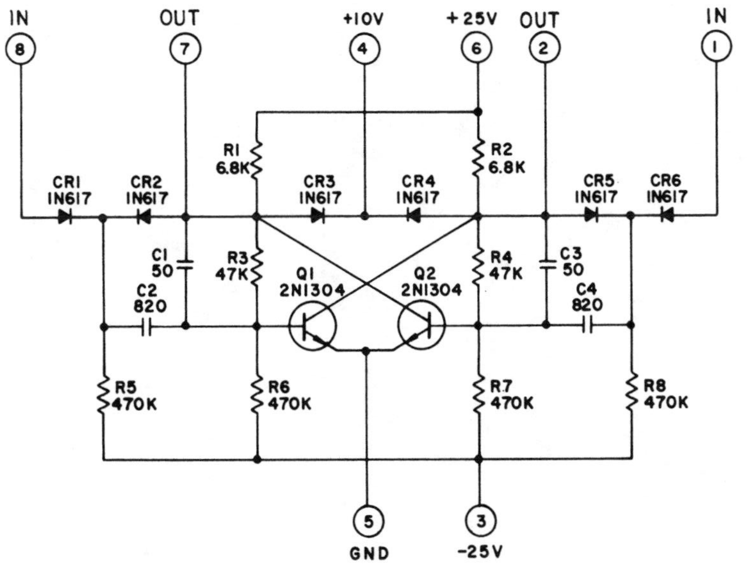

IN 8, OUT 7, +10V 4, +25V 6, OUT 2, IN 1

R1 6.8K, R2 6.8K, CR1 IN617, CR2 IN617, CR3 IN617, CR4 IN617, CR5 IN617, CR6 IN617

C1 50, C2 820, R3 47K, Q1 2N1304, Q2 2N1304, R4 47K, C3 50, C4 820

R5 470K, R6 470K, R7 470K, R8 470K

5 GND, 3 −25V

NOTES:

1. ALL RESISTANCE VALUES ARE IN OHMS, (K = 1000, M = 1000000) ¼W, 5%.
2. ALL CAPACITANCE VALUES ARE IN MICROMICROFARADS.

Exercise: PCB-5

Start with the Layout below (2 × size). Make the length of the board layout as compact as possible. Spacing between components, land, conductors, etc. should be $\frac{1}{32}$ minimum ($\frac{1}{16}$ in 2 × size layout). All required data may be found in Appendixes (outline and hardware) B and C. Identify this board as PCB-5.

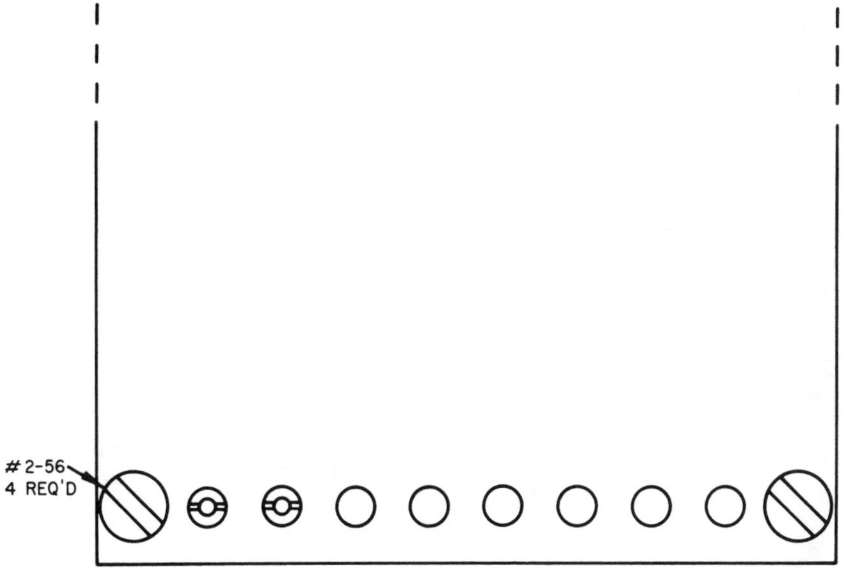

#2-56
4 REQ'D

NOTE:
Terminals need not be numbered in sequential order.

TITLE PRINTED CIRCUIT BOARD–FLIP-FLOP (SINGLE SIDED)					DWG. NO. PCB-5	
NAME	DATE	COURSE	GRADE	SCALE 2 X SIZE	SHEET 1 OF 3	PAGE 48

Exercise: PCB-5 (MASTER, ARTWORK)
Complete the master of PCB-5 below. Use either pencil or tapes, as instructed by your teacher.

Exercise: **PCB-5** (DRILLED)
Complete the drilled board of PCB-5 below (see PCB 4).

TITLE PRINTED CIRCUIT BOARD—FLIP-FLOP (SINGLE SIDED)					DWG. NO. PCB-5	
NAME	DATE	COURSE	GRADE	SCALE 2 X SIZE	SHEET 2 OF 3	PAGE 49

Exercise: PCB-5 (ASSEMBLY)
Complete the assembly and list of material of PCB-5 below.

LIST OF MATERIAL

ITEM	NO. REQD.	REFERENCE DESIGNATION	DESCRIPTION	MANUFACTURER AND PART NUMBER OR MIL — TYPE DESIGNATION
1	X		SCHEMATIC	SHEET 1 OF 3
2		PCB-5		
3		C1, C3		
8				
		R5, R6, R7, R8		

TITLE PRINTED CIRCUIT BOARD – FLIP-FLOP (SINGLE SIDED)					DWG. NO. PCB-5	
NAME	DATE	COURSE	GRADE	SCALE 2 X SIZE	SHEET 3 OF 3	PAGE 50

PRINTED CIRCUIT BOARD PCB-6.

From the preceding schematic and the same list of material of PCB-5, prepare all the drawings required (four more) to fabricate a **double-sided PCB-6** (see PCB-4).

Exercise: (PCB-6)

Start with the layout below (2 × size). Make the length of the board layout as compact as possible, smaller than PCB-5. Use tabs in this fabrication so that the board can be plugged into a connector. No mounting screws are required. Spacing between components, land, conductors, etc., should be $\frac{1}{32}$ minimum ($\frac{1}{16}$ in 2 × size layout). Use the same data as in the preceding exercise (PCB-5). Identify this board as PCB-6.

(**Hint:** Two-sided boards require tool holes.)

A TYPICAL PLUG-IN
PRINTED CIRCUIT BOARD

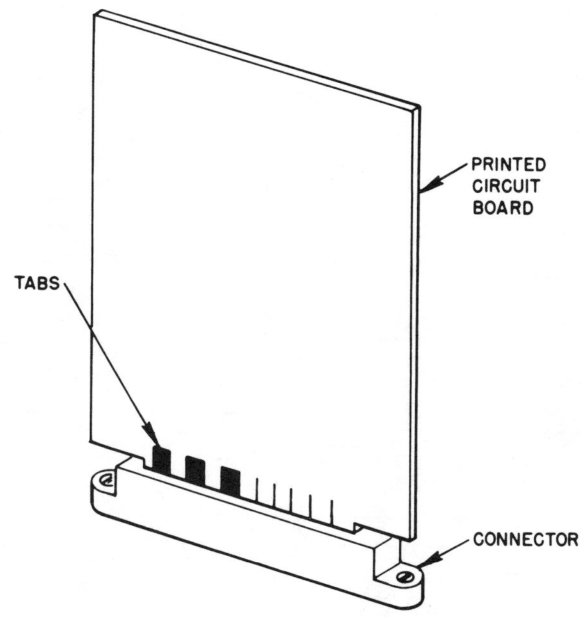

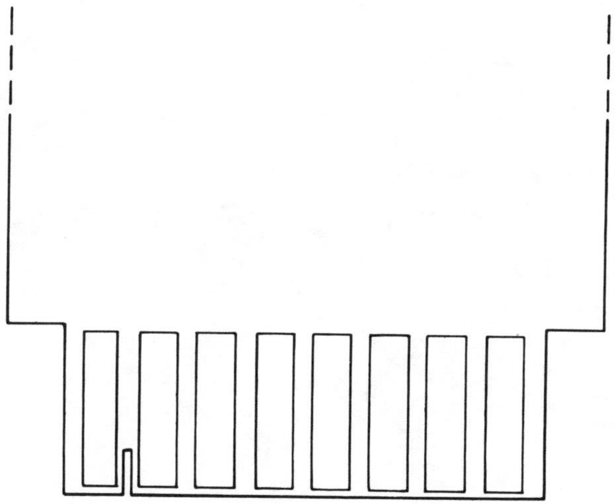

TABS

PRINTED
CIRCUIT
BOARD

CONNECTOR

TITLE PRINTED CIRCUIT BOARD–FLIP-FLOP (DOUBLE SIDED)					DWG. NO. PCB–6	
NAME	DATE	COURSE	GRADE	SCALE 2 X SIZE	SHEET 1 OF 4	PAGE 51

Exercise: PCB-6 (MASTER, ARTWORK)
Complete the two masters of PCB-6 below. Use either pencil or tapes, as instructed by your teacher.

FRONT FACE

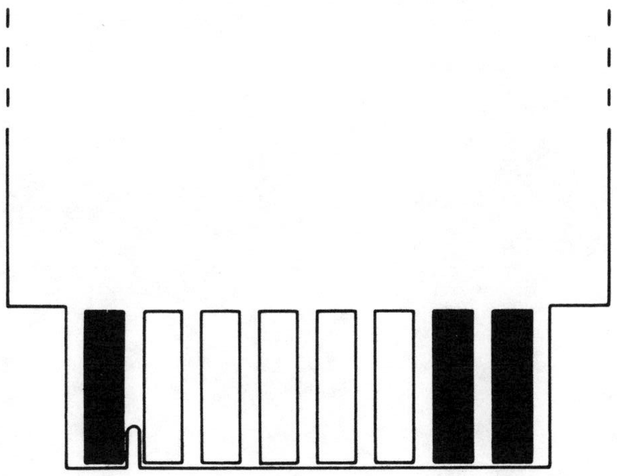

BACK FACE

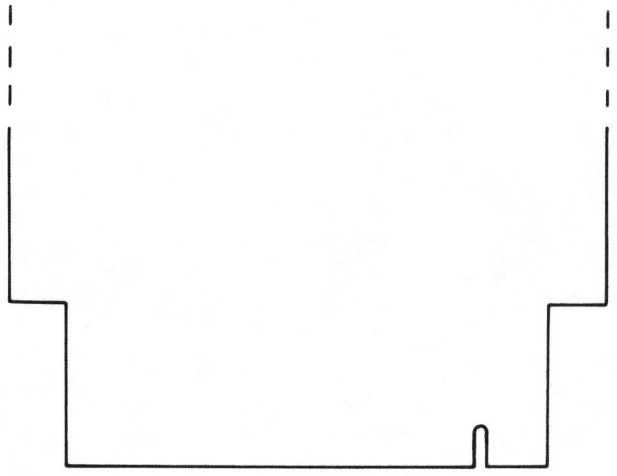

TITLE PRINTED CIRCUIT BOARD—FLIP-FLOP (DOUBLE SIDED)					DWG. NO. PCB-6	
NAME	DATE	COURSE	GRADE	SCALE 2 X SIZE	SHEET 2 OF 4	PAGE 52

Exercise: PCB-6 (DRILLED)

Complete the drilled board of PCB-6 below. Show all notes and hole chart (See PCB-4).

TITLE PRINTED CIRCUIT BOARD — FLIP-FLOP (DOUBLE SIDED)					DWG. NO. PCB-6	
NAME	DATE	COURSE	GRADE	SCALE	SHEET 3 OF 4	PAGE 53

Exercise: PCB-6 (ASSEMBLY)

Complete the assembly of both sides (front and back) of PCB-6 below. Since this exercise has the same list of materials as PCB-5, it is not necessary to show it again. Draw only the components and outline of boards; do not show lands or conductors.

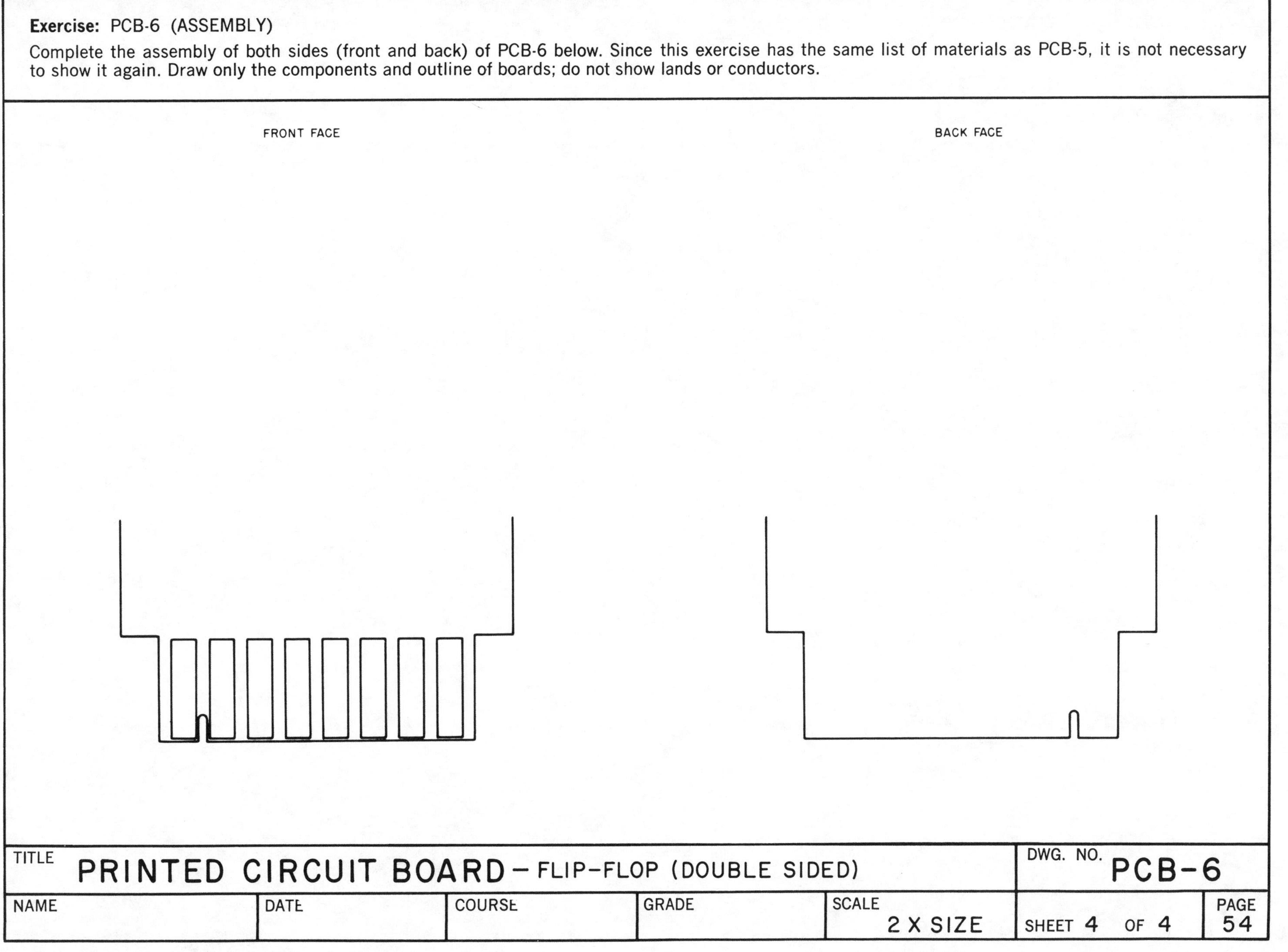

TITLE PRINTED CIRCUIT BOARD – FLIP-FLOP (DOUBLE SIDED)					DWG. NO. PCB-6	
NAME	DATE	COURSE	GRADE	SCALE 2 X SIZE	SHEET 4 OF 4	PAGE 54

INTRODUCTION: ELECTROMECHANICAL DESIGN

Basic steps needed in Electromechanical design are:

1. **Schematic diagram** (given to you in sketch form)
2. **Preliminary electrical parts list** (given)
3. **Layout** (done by you as designer)
4. **Detail drawings** (mechanical, from catalogues and/or direct measurements)
5. **Final assembly drawing** and **complete list of material**
6. **Wiring diagram** (both steps 5 and 6 done by you)

POWER SUPPLY DESIGN

Exercise. An electronics engineer wants you to design a simple package of the power supply given in the schematic shown in Fig. 1.

Solution. The first thing to do is to study the schematic. Then make a **preliminary electrical parts list** and **outline dimensions** of all the **components** given in the schematic.

Complete the preliminary electrical parts list by selecting the proper electrical components from Appendix B.

Next, devise a simple, neat, easy-to-build, serviceable package (continued on the next pages).

Note: XF1 is related to F1 by the following MIL-STD definition: A socket, fuseholder, or similar device that is always associated with a single particular part or subassembly (such as an electron tube, a fuse etc.) shall be identified by a composite reference designation consisting of the class letter "X," followed by the basic reference designation that identifies the mounted part. For example, the socket for fuse F1 would be XF1.

POWER SUPPLY SCHEMATIC DIAGRAM

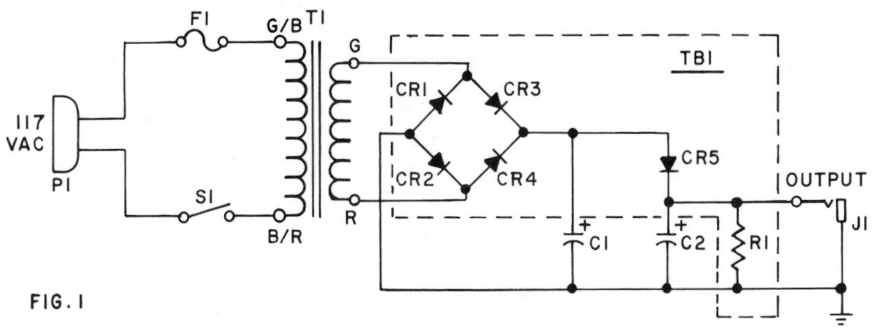

FIG. 1

Given:

Capacitor, C1 & C2 = 1500mfd, 50V
Diode, CR1 Thru CR5 = 1N91
Resistor, R1 = 22K, ½W, 5%
Transformer, T1 = TRIAD No. F-92A
Fuse, F1 = 1 amp, 125V (3 AG)
Fuse Holder, XF1 = Littelfuse, No. 342001
Toggle switch, S1 = MS35058-22
Output Jack, J1 = Pin Jack Strip (two pins)
Power cord, P1 = any standard cord (black)

PRELIMINARY ELECTRICAL PARTS LIST AND OUTLINE DIMENSIONS OF COMPONENTS

See Appendix B and complete the list below:

C1 and C2 = 1⅜ dia. × 3$\frac{9}{16}$ high (ARCO)

CR1 through CR5 — ____________

P1 = ¼ O.D. cord 6 FEET LONG, STOCK ITEM

R1 = ____________ (RC20GF223J)

T1 = ____________

XF1 = ____________

F1 = ____________ (LITTELFUSE, INC.)

S1 = ____________

J1 =

TITLE ELECTROMECHANICAL DESIGN-INTRODUCTION					DWG. NO. EMD-1	
NAME	DATE	COURSE	GRADE	SCALE	SHEET 1 OF 2	PAGE 55

ELECTROMECHANICAL DESIGN (cont.)

From your **preliminary electrical parts list** you will notice that transformer (T1) and capacitors (C1 and C2) are the largest components. One must select an enclosure or chassis large enough to house all the components and necessary hardware. This brings us to the third step — namely, the **layout.**

LAYOUT

The layout should describe the parts sufficiently for the detailer or a draftsman to understand what has been selected. Common screws or any other hardware may be written on the layout near the part for clarity.

The layout should be drawn accurately to scale. In this case we will use half scale. Other accessories which will be needed in our package will include the following:

*1. **Terminal board** (TB1) for mounting all diodes and resistor.

2. **Fuse holder** to house fuse (F1).

3. **Stand-offs, grommet, terminal lug, screws, nuts, washers,** etc.

You will find that a **chassis** 5½ × 4¾ × 1⁹⁄₁₆ would be large enough for our purpose.

Exercise. Complete the two views of the started layout on the right (scale: half size). Identify all components with reference designations. See Schematic.(Page 55)

In general, additional information required for the design and/or details (such as **material for the chassis, finishes, specs,** etc.) are given in note form.

*Make terminal board (TBI) 3.50 inches long with CR1, CR2, CR3, and CR4 mounted in the same manner as R1 and CR5.

CHASSIS LAYOUT.

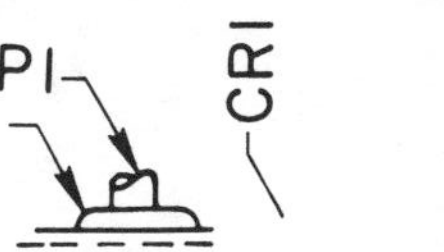

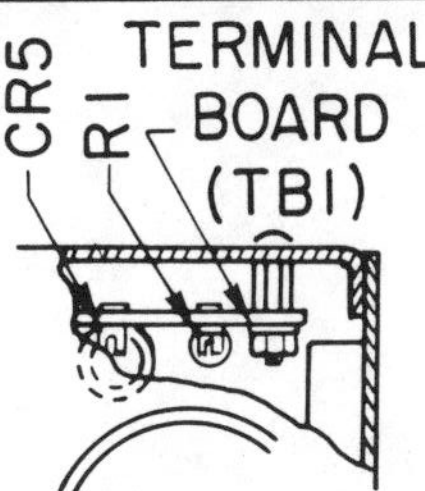

CHASSIS

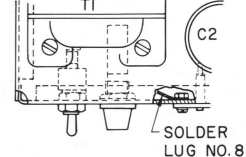

TOP VIEW

TI

C2

SOLDER
LUG NO. 8

NOTES:

1. Material: alum sheet, 1⁄16 thk. 6061-T6, per QQ-A-327, cond T.
2. Finish: alodine per MIL-C-5541 external surfaces and edges shall be painted gray.
3. Silk screen or rubber stamp letters on front, black.
4. No. 8-32 screws are used on T1 and J1 (4 and 2 req'd). No. 4-40 screws and stand-offs are used on TB1 (4 req'd).

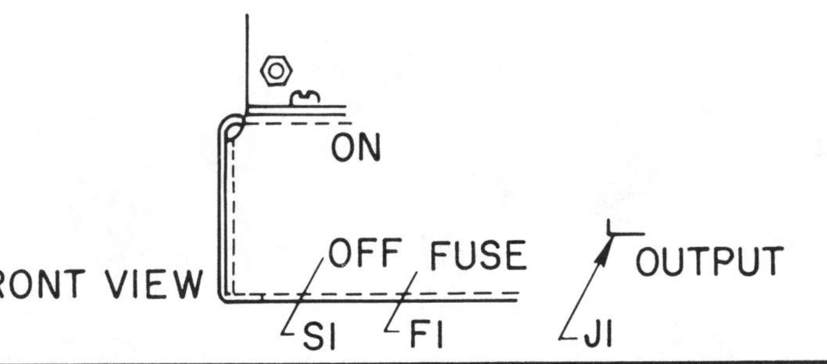

FRONT VIEW

TITLE ELECTROMECHANICAL DESIGN—LAYOUT					DWG. NO. EMD-1	
NAME	DATE	COURSE	GRADE	SCALE 1/2	SHEET 2 OF 2	PAGE 56

ELECTROMECHANICAL DESIGN (cont.)

From the layout we proceed to the fourth step — namely, the **details.** The details are drawn by a draftsman or detailer from the layout. Detail parts should show dimensions, tolerance, etc. Whenever possible, for clarity, show the detail as large as space and standard scaling permit (e.g., 2/1 or 4/1). Use as many views and sections as are necessary to completely describe the part. It would be a good idea to detail the **terminal board** (TB1) before you detail the chassis.

Exercise: TERMINAL BOARD (TB1) DETAIL

From the layout you will notice that six components are mounted on the terminal board: five **diodes** and one **resistor.** That means that twelve **terminals** are required (two terminals per component).

Draw the terminal board full scale. Terminals are USECO No. 2000 B. (See Hardware, Appendix C, page 123). Show the board, terminals, and markings all in one detail.

Board dimensions are 1.40 × 3.50 × .062 thick.

Terminal spacings are .50 horizontally and 1.00 vertically.

To determine BD MTG hole callout, assume #4-40 screws to be used with .005 ℄ - ℄ tol. and clearance on clearance hole mtg (see Appendix D, page 128).

For Terminals — Be sure to use a callout for terminals and their mtg holes.

Dimension the terminal board according to standard practice as in Lesson 1, (Mechanical Drafting Review). Silk Screen or rubber stamp terminals from 1 through 12 as shown. Identify board TB1 parallel to right edge. In practical applications the drilling and marking details are done separately, but both details are combined here for simplicity.

After completing the details, proceed to the next page for **terminal board assembly.**

Exercise. Complete the two views of the terminal board below.

.15
1
7
.50
.062

NOTES:

1. TERMINAL BOARD MATERIAL: .062 THK. PHENOLIC
2. SILK SCREEN OR RUBBER STAMP NUMBERS AND TERMINAL BOARD IDENTIFICATION (TBI) ⅛ HIGH, BLACK, BEFORE INSTALLING TERMINALS.
3. INSERT TERMINALS INTO BOARD AND SWAGE THEM OVER.

TITLE ELECTROMECHANICAL DESIGN-TERMINAL BOARD DETAIL					DWG. NO. EMD-2	
NAME	DATE	COURSE	GRADE	SCALE FULL	SHEET 1 OF 2	PAGE 57

TERMINAL BOARD ASSEMBLY

The **assembly board** is drawn in the same manner as has been done in previous lessons. The only difference is that the Assembly here is drawn in two stages. This is sometimes called a **stage drawing.**

Stage 1. In this step, only the **board** and the **jumper wires** are shown. The terminals are connected with jumper wires. You will have to check this against the schematic and layout to determine which terminal will require the jumpers. Also show wires coming out of the board and identify the routes of these wires by assigning a **color code** to each. The color code will later help the technician to complete the wiring of the **power supply.** There are three wires coming out of the board:

From junction CR1, CR2, and R1, one wire (black)
From junction CR3, CR4, and CR5, one wire (red)
From junction R1 and CR5, one wire (green)

These wires should be at least 10″ long.

Stage 2. Draw the board again, but in this view show only the **electrical components** (CR1 through CR5 and R1). Show reference designations of all components. They should conform to those on the schematic diagram. With the assembly, include a list of material.

Note: If this board had been made into a **printed circuit board,** stage 1 would have been omitted. Due to the simplicity of the board, wires were used.

Complete the exercises on the right.

Exercise. Complete two stages and the list of material below.

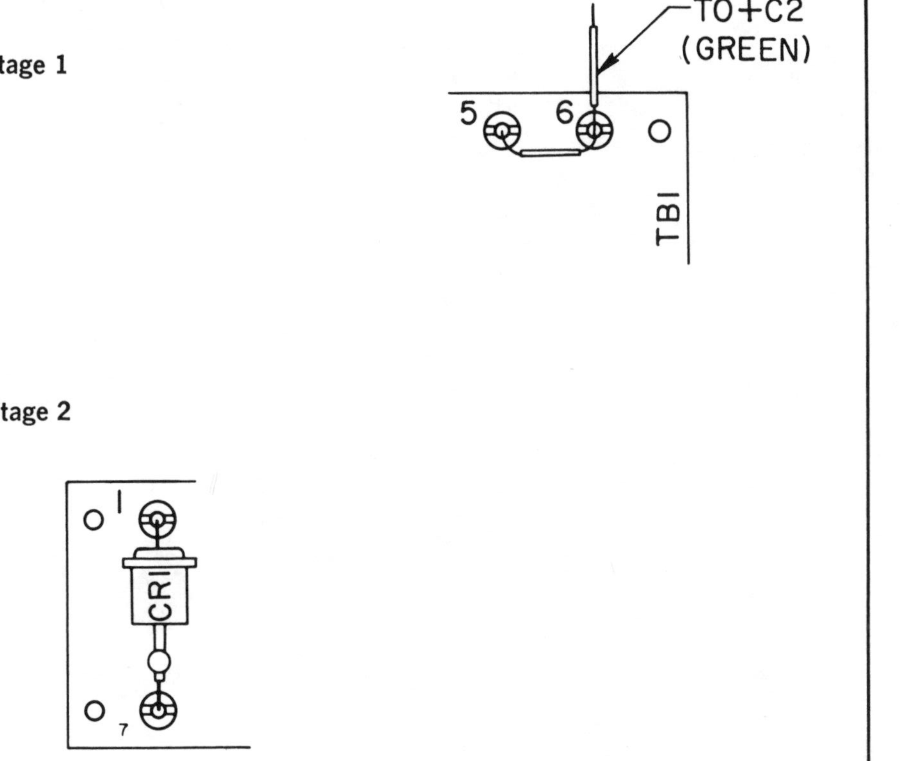

	AR		WIRE	AWG.22, PER MIL-W-16878
	AR		SOLDER	PER QQ-S-571
	1		RESISTOR	
2		CRI-CR5		
1		TBI		
ITEM	NO. REQ'D	REF. DESIG.	DESCRIPTION	MANUFACTURER AND PART NO. OR MIL TYPE DESIGNATION

LIST OF MATERIAL

TITLE ELECTROMECHANICAL DESIGN—TERM. BD. ASS'Y—STAGE DRAWINGS					DWG. NO. EMD-2	
NAME	DATE	COURSE	GRADE	SCALE FULL	SHEET 2 OF 2	PAGE 58

Exercise. Complete the drawing of the **chassis** below showing three views (front, top, and back) plus Detail H (hole pattern for C1 and C2), and fill in **hole chart.** (See Lesson #1 for reviewing hole-pattern details). Scale: half size

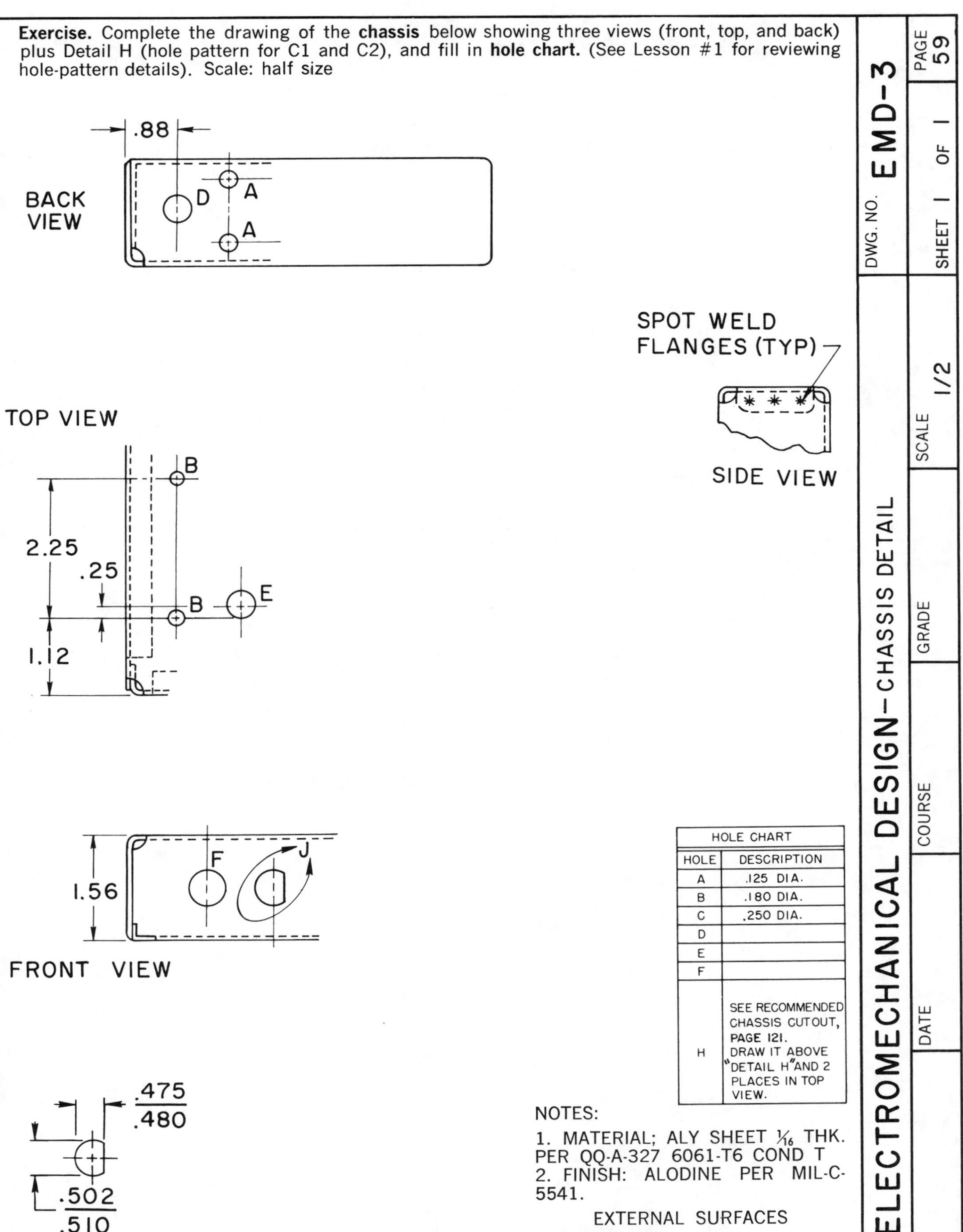

HOLE CHART	
HOLE	DESCRIPTION
A	.125 DIA.
B	.180 DIA.
C	.250 DIA.
D	
E	
F	
H	SEE RECOMMENDED CHASSIS CUTOUT, PAGE 121. DRAW IT ABOVE "DETAIL H" AND 2 PLACES IN TOP VIEW.

NOTES:

1. MATERIAL; ALY SHEET 1/16 THK. PER QQ-A-327 6061-T6 COND T
2. FINISH: ALODINE PER MIL-C-5541.

EXTERNAL SURFACES TO BE PAINTED GREY.

SILK SCREEN OR RUBBER STAMP ALL IDENTIFICATIONS (ON, OFF, FUSE, OUTPUT)

TITLE ELECTROMECHANICAL DESIGN–CHASSIS DETAIL

DWG. NO. EMD-3 | PAGE 59 | SHEET 1 OF 1 | SCALE 1/2 | GRADE | COURSE | DATE | NAME

ASSEMBLY DRAWINGS

Assembly drawings should be drawn so that the detailed parts and **sub-assemblies** are shown in their relative position and scale with the number of views necessary to clearly portray the proper attachments. All detailed parts and sub-assemblies should be identified by an **item number** and, if an electrical part, by marking its **circuit designation** on the component. The item number should be placed in a $\frac{3}{8}$ diameter circle with a lead running from the circle to the part it identifies.

LIST OF MATERIAL

For each assembly drawing prepare a **list of material** and an **electrical parts list** containing all items which become part of the completed assembly, including (1), every item purchased or fabricated and (2), all material including finishes.

A subassembly which becomes part of the assembly should be listed as one item and identified by its drawing number (for example, the terminal board TB1). The list of material (hardware and fabricated parts) and the list of electrical parts should be prepared separately.

Exercise. Complete the **assembly drawing** on the right. If you made a good layout it could be used as the assembly drawing (see EMD-1, page 56).

Show all items and identify the electrical components with **reference designations.** The reference designations of all electrical components should conform to those on the schematic diagram.

On the next page prepare separately, a list of material and an electrical parts list. For references use the schematic diagram, layout, and the appendixes.

The wiring diagram of the "power supply" will be drawn in Lesson 10 on wiring diagrams.

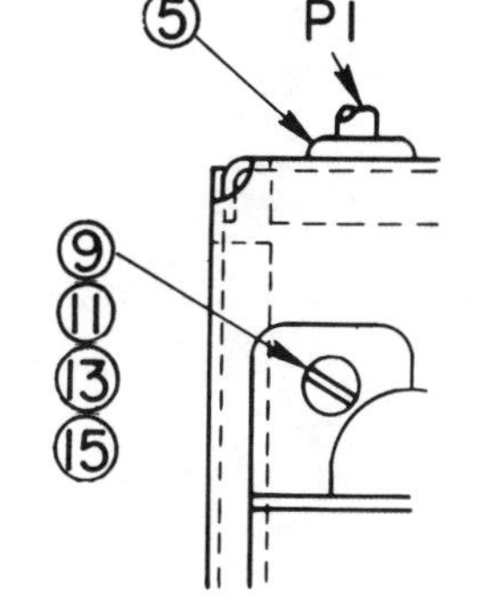

TITLE ELECTROMECHANICAL DESIGN—ASSEMBLY DRAWING					DWG. NO. EMD-4	
NAME	DATE	COURSE	GRADE	SCALE 1/2	SHEET 1 OF 2	PAGE 60

Complete the List of Material and the Electrical Parts List below. Use the assembly drawing and appendixes for reference.

ITEM	NO. REQ'D	REF. DESIG	DESCRIPTION	MANUFACTURER AND PART NO. OR MIL TYPE DESIGNATION
10	AR		HOOKUP WIRE	AWG. 22, PER MIL-W-16878
9	AR		SOLDER SOFT	PER QQ-S-571
8		T1		TRIAD TRANSFORMER CORPORATION NO. F-92A
7	1		SWITCH TOGGLE	
6	1		PLUG AC	2 CONDUCTOR CORD, RUBBER-JACKETED AWG 18 (41 × 34) .245 O.D.
5	1	J1	JACK	
4			CAPACITOR ELECTROLYTIC	
3	1		FUSE	
2	X		SCHEMATIC DIAGRAM	EMD-1 Sheet 1 of 1
1	X		WIRING DIAGRAM	WD-2 Sheet 1 of 1 (Page 63)

ELECTRICAL PARTS LIST

ITEM	NO. REQ'D	DESCRIPTION	MANUFACTURER & PART NO. OR MIL TYPE DESIGNATION
15	6	WASHER, LOCK, SPLIT, NO. 8	
14	4	WASHER,	MS35337-78
13	6	WASHER,	MS15795-307
12	4	WASHER, FLAT, NO. 4, CRES	
11			MS35649-84
10	4		MS35649-44
9	4	SCREW, PAN HD. NO. 8-32 × 3/8 LONG	
8	2		MS35221-44
7	4		MS35221-18
6	1	SOLDER LUG (NO. 8)	
5	1		
4			
3	1		LITTELFUSE, INC. NO. 342001
2	1	TERMINAL BOARD ASSEMBLY (TB1)	EMD-2 SHEET 2
1		CHASSIS	EMD-3 SHEET 1

LIST OF MATERIAL

TITLE ELECTROMECHANICAL DESIGN-ELECTRICAL & MECHANICAL PARTS LIST

DWG. NO. EMD-4

SHEET 2 OF 2

SCALE

GRADE

COURSE

DATE

NAME

INTRODUCTION: THE WIRING DIAGRAM

A **wiring diagram** (or connection diagram) shows pictorially, or in list form, the **wire connections** of an electronic assembly or its components. These connections may be external, internal, or both; however, the external-connection drawing is usually referred to as an **interconnection diagram.** An example shown in Fig. 1.

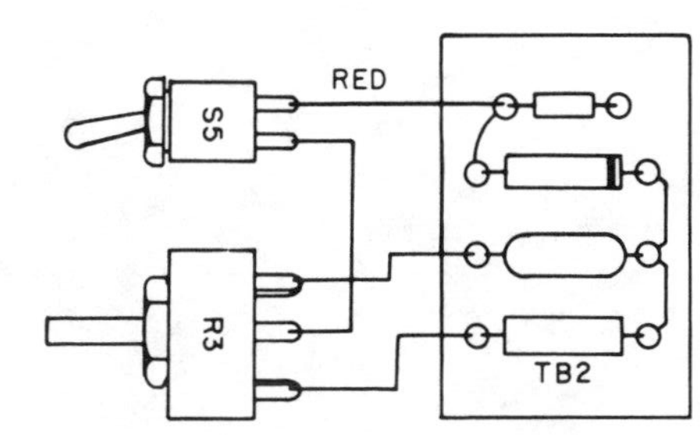

FIG. 1

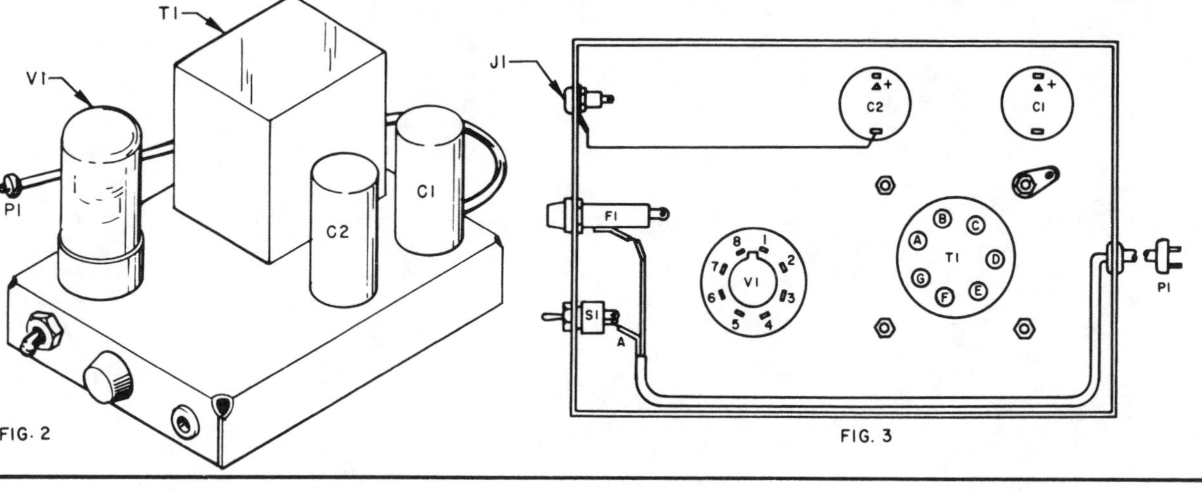

FIG. 2

FIG. 3

Exercise. Shown in Fig. 2 is the top view of the **power supply** in lesson 5, SCH-1 sheet 1 (page 25) Fig. 3 is the bottom view of the power supply.

Redraw the bottom view of the power supply and show the wiring connections between all the components. For reference see schematic diagram of the power supply in SCH-1.

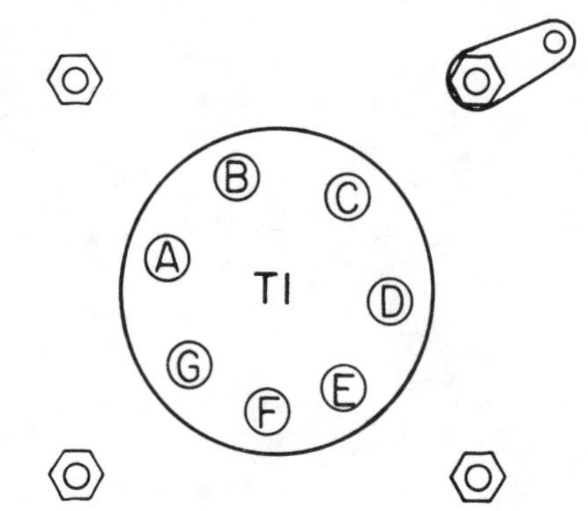

TITLE WIRING DIAGRAM–INTRODUCTION					DWG. NO. WD-1	
NAME	DATE	COURSE	GRADE	SCALE NONE	SHEET 1 OF 1	PAGE 62

Exercise. Complete the started wiring diagram and wiring list of the power supply that you designed in the lesson on electromechanical design. Reference designations of the components should conform to those in the schematic diagram of the power supply (EMD-1, sheet 1, page 55). Notice that **blocks** are being used to represent component outlines. The **wiring list,** which is self-explanatory here, is commonly used in very complicated wiring diagrams.

117 VAC

P1

F1

S1

T1

G/B

G

B/R

R

TB1

12 11 10 9 8 7 6 5 4 3 2 1

C1

C2

J1

OUTPUT

GROUND LUG

WIRE			ITEM NO.	FROM	TO
LENGTH INCHES	AWG	COLOR		LOCATION	LOCATION
6	22	BLACK	1	P1	F1
			2		S1
		G/B	3	T1	F1
			4	T1	
9		G	5	T1	TB1-1
		R	6	T1	
		G	7	TB1-6	
		RED	8	TB1-11	+ C1
			9	TB1-12	
		BLACK	10	C1	
		BLACK	11	C2	
			12	J1	GND. LUG
	22		13	+ C2	

ALL WIRES
#22 AWG.

TITLE WIRING DIAGRAM—POWER SUPPLY					DWG. NO. WD-2	
NAME	DATE	COURSE	GRADE	SCALE NONE	SHEET 1 OF 1	PAGE 63

INTRODUCTION: THE INTERCONNECTION DIAGRAM

An **interconnection diagram** is a drawing which shows the external wiring connections between items of equipment or unit assemblies. The internal connections of the unit assemblies are generally omitted.

In Fig. 1 is shown a typical audio system consisting of five units:

UNIT NO. 1: Record player
UNIT NO. 2: Tape recorder
UNIT NO. 3: Preamp
UNIT NO. 4: Power amplifier
UNIT NO. 5: Speaker

All these units are interconnected with cables numbered from W1 through W4. This audio system can be shown drawn as an **interconnection diagram** by drawing each unit as a **block** and each cable as a line, thus eliminating pictorial drawings.

A **unit** is identified by a unit number such as 1, 2, 3, etc.

A **cable** is identified with a W such as W1, W2, W3, etc.

The unit will always show the connector (jack) marked J1, J2, etc. Thus 2J1 means UNIT NO. 2, JACK NO. J1. The cables will be identified as shown:

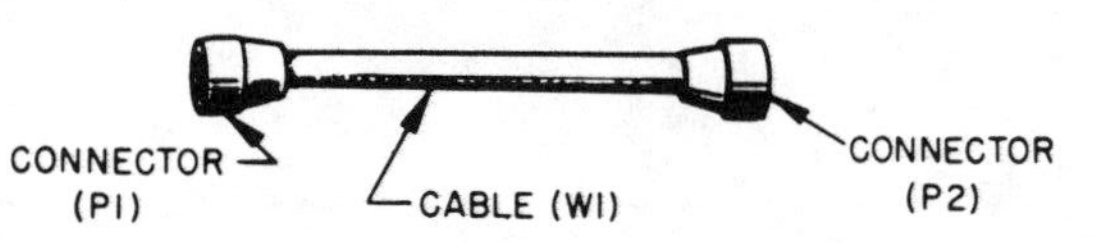

Thus W1P1 means CABLE NO. W1, CONNECTOR NO. P1.

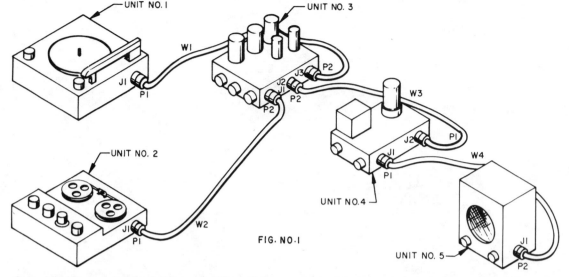

FIG. NO. 1

Exercise. Complete the **interconnection diagram** started below of the audio system (Fig. 1) by showing a single-line drawing identifying all **blocks, units, connectors,** and **cables.**

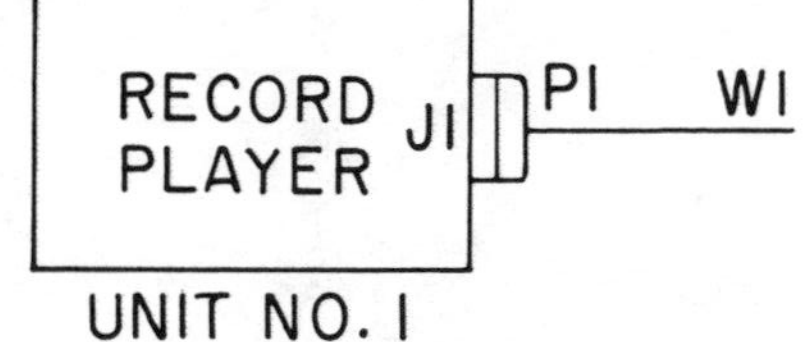

TITLE INTERCONNECTION DIAGRAM–AUDIO SYSTEM					DWG. NO. ID–1	
NAME	DATE	COURSE	GRADE	SCALE NONE	SHEET 1 OF 1	PAGE 64

Exercise. Draw an **interconnection diagram** of a digital computer system and a list of routing of all cables without the help of a sketch. In more complicated interconnection diagrams, a list of cable routing is extremely helpful. The digital computer consists of the following nine units:

- UNIT NO. 1: Arithmetic unit (6 jacks, J1 through J6)
- UNIT NO. 2: Typewriter (1 jack, J1)
- UNIT NO. 3: Tape recorder (1 jack, J1)
- UNIT NO. 4: Card reader (1 jack, J1)
- UNIT NO. 5: Plotter (1 jack, J1)
- UNIT NO. 6: Power supply (1 jack, J1)
- UNIT NO. 7: Core memory (1 jack, J1)
- UNIT NO. 8: Disk storage (1 jack, J1)
- UNIT NO. 9: Tape storage (1 jack, J1)

All the **units** from UNIT NO. 2 through 9 have one single **jack** (J1). UNIT NO. 1 has 6 JACKS numbered from J1 through J6 as shown. Show cable routing as follows;

From UNIT NO. 1 to UNIT NO. 2 J1 (CABLE NO. W1)
UNIT NO. 3 J1 (CABLE NO. W2)
UNIT NO. 4 J1 (CABLE NO. W3)
UNIT NO. 5 J1 (CABLE NO. W4)
UNIT NO. 6 J1 (CABLE NO. W5)
UNIT NO. 7 J1
UNIT NO. 8 J1
From UNIT NO. 1 to UNIT NO. 9 J1 (ONE SINGLE CABLE W6)

Complete the CABLE ROUTING LIST.

CABLE ROUTING

FROM	TO
W1P1	2J1
W1P2	1J1
W6P4	7J1

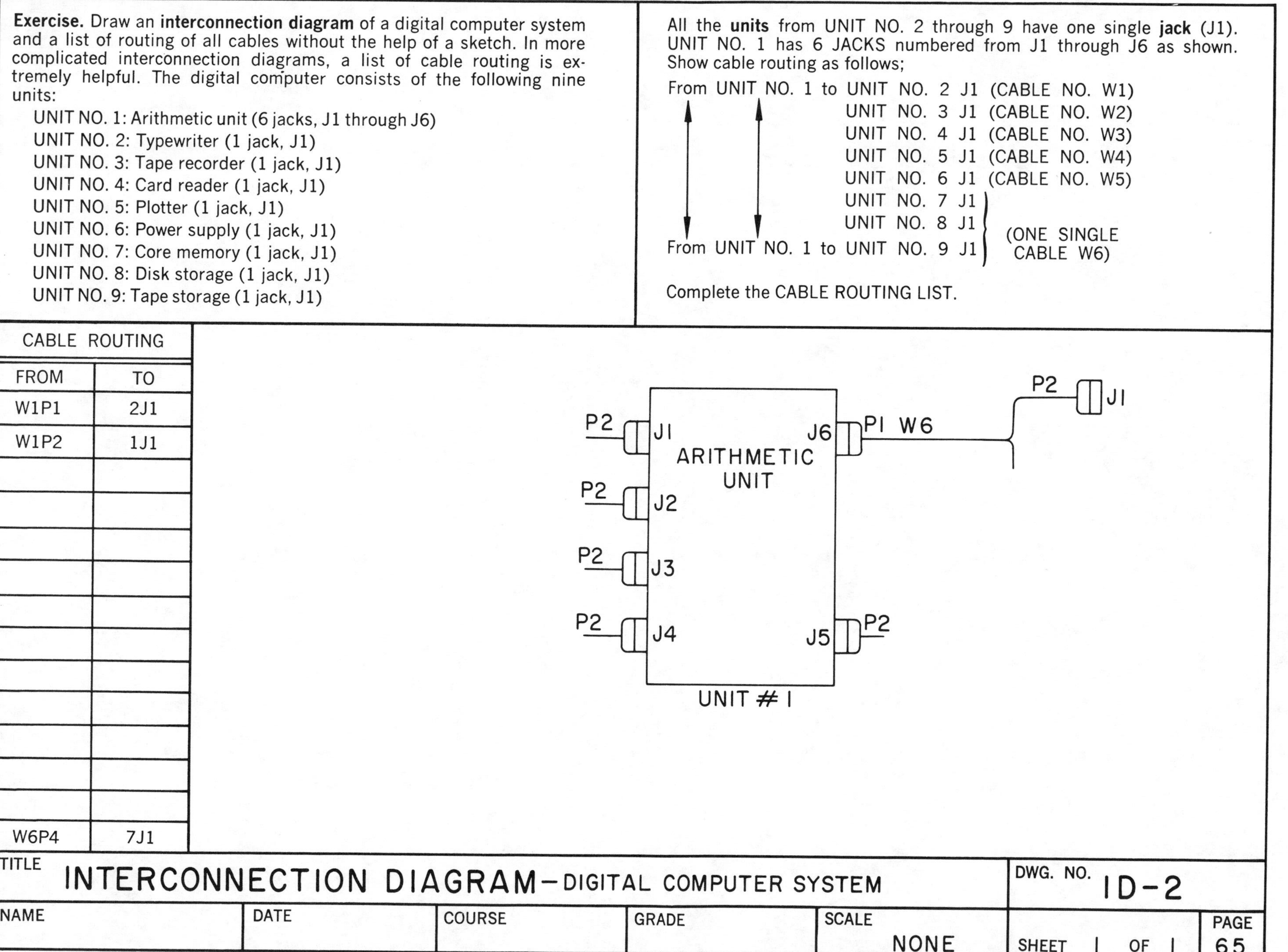

TITLE INTERCONNECTION DIAGRAM—DIGITAL COMPUTER SYSTEM

DWG. NO. ID-2

NAME	DATE	COURSE	GRADE	SCALE NONE	SHEET 1 OF 1	PAGE 65

PICTORIALS

In most large companies, a technical illustrator does the "formal" **pictorials.** Smaller companies have their own draftsmen do them. But perhaps the most widespread function of the pictorial is its use in sketches between draftsmen, designers, and engineers to develop ideas. There are many types of pictorials. Probably the most common and easiest to draw are the **cabinet** and **isometric drawings.** These are illustrated at right. Neither the cabinet nor the isometric drawing is an accurate picture. Therefore a certain amount of "fudging" (guesswork and inaccurate touch-up) is acceptable.

Exercise 1. Draw a second bracket on cabinet and isometric views (top and middle).

Exercise 2. In the isometric boxes at bottom right draw three 8-point circles in each surface center of box A and three 4-point center circles in each surface center of box B. Make all holes ½″ diameter.

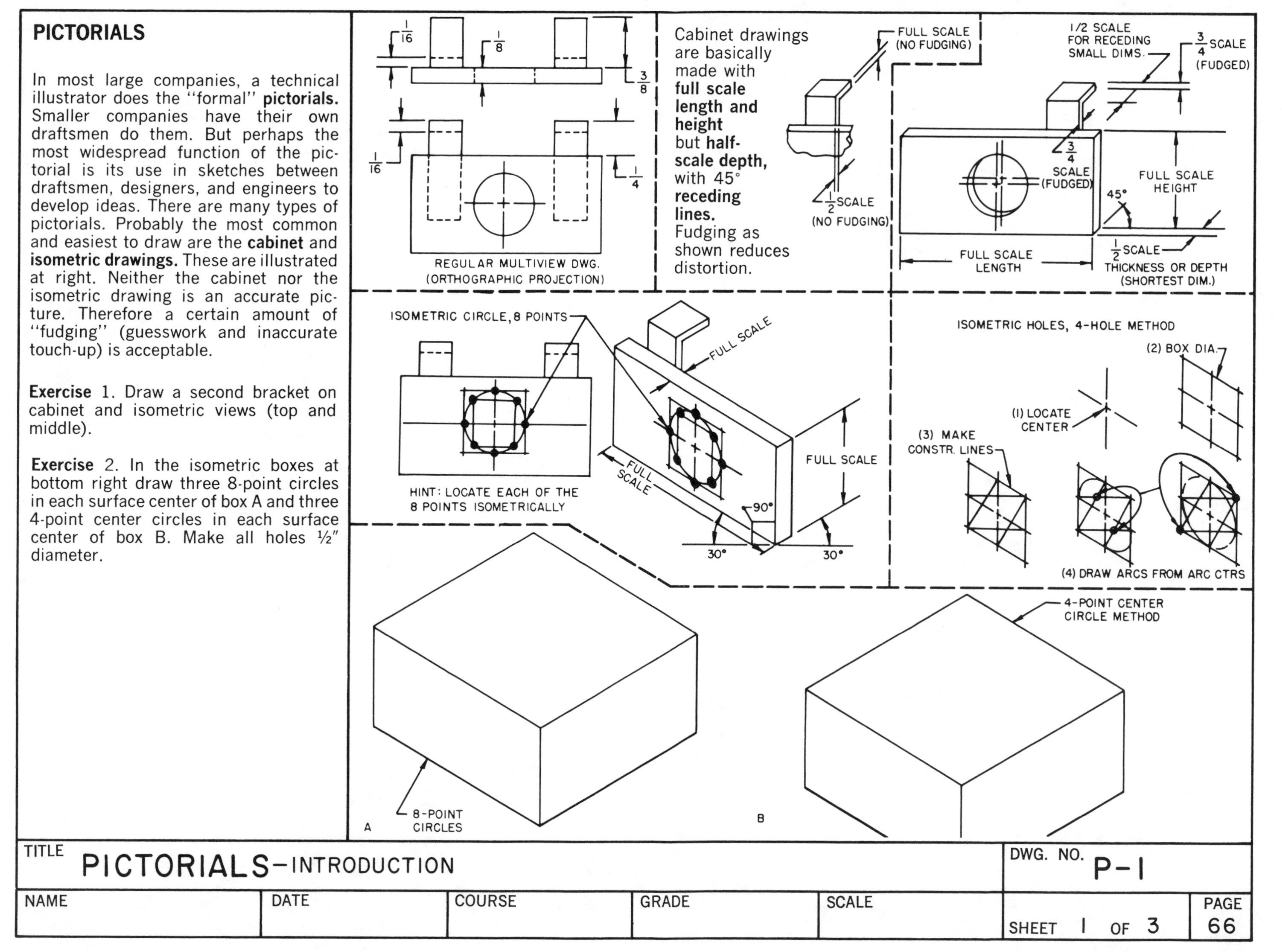

TITLE PICTORIALS–INTRODUCTION					DWG. NO. P–1	
NAME	DATE	COURSE	GRADE	SCALE	SHEET 1 OF 3	PAGE 66

From sheet 1 it was apparent that **cabinet drawings** lend themselves to relatively thin objects, while the boxlike subject shows up better in **isometric drawings.**

Exercise. Complete the cabinet drawing started below of the **terminal board assembly,** 2 X size. For reference see Electromechanical Design (lesson 9) sheets 1 and 2, pages 57 and 58. The terminals are USECO No. 2000B (Appendix C, page 123).

TITLE PICTORIALS–CABINET DRAWING–TERMINAL BOARD					DWG. NO. P-1	
NAME	DATE	COURSE	GRADE	SCALE 2/1	SHEET 2 OF 3	PAGE 67

In **isometric drawings,** a cylinder can be drawn like a hole. Any curved line can be regarded as a portion of a circle and may be located by a series of points. In using a template (ellipse) for ISOMETRIC CIRCLES or curves, be sure to properly align the **major axis** of the ellipse so that it is 90° or at right angles to the **hole axis** (Fig. 1).

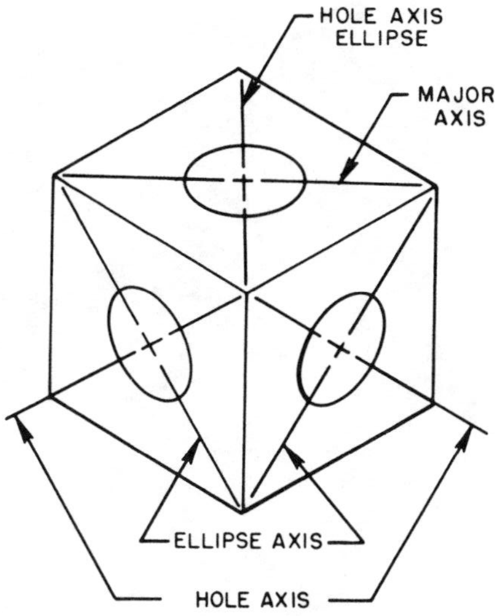

HOLE AXIS ELLIPSE
MAJOR AXIS
ELLIPSE AXIS
HOLE AXIS

Fig. 1

Exercise. Complete the **isometric** below. Draw the power supply of lesson 9, EMD-4 sheet 1 (page 60).

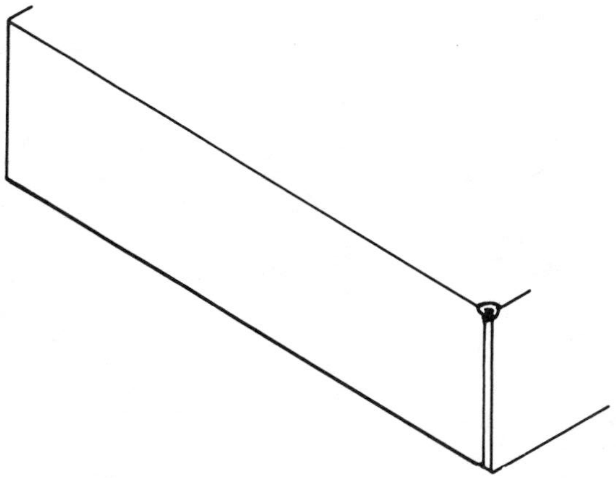

SCALE: HALF SIZE

TITLE PICTORIALS—ISOMETRIC DRAWING					DWG. NO. P-1	
NAME	DATE	COURSE	GRADE	SCALE 1/2	SHEET 3 OF 3	PAGE 68

Note: Three-Quarter size template MIL-STD-806 will be required for the next exercises (page 70).

LOGIC TEMPLATE. The standard logic symbol template No. MIL-STD-806 shown at right is used to draw logic symbols in much the same manner as the ANSI Y32.2 template is used to draw Electronic Schematic Symbols.

The MIL-STD-806 is often referred to as ASA Y32.14 since they are one and the same.

To familiarize yourself with the symbols, study Appendix F (pages 130 and 131) and do the exercises below.

LOGIC TEMPLATE MIL-STD-806, THREE-QUARTER SIZE

STANDARD LOGIC SYMBOLS
THREE-QUARTER SIZE

MIL-STD-806
ASA Y32.14

1 What is a flip-flop? (See explanation, page 130.)

THE FLIP-FLOP IS ______

______

______

2 Complete the statement: "THE OR OUTPUT IS HIGH (H) IF AND ONLY IF (page 130) ______

______

3 If the INCLUSIVE OR inputs were: A = **HIGH**, B = **LOW**, C = **LOW**; THE OUTPUT F WOULD BE ______

4 If the INCLUSIVE OR function above (Problem 3) had a **small circle** attached to the output, the output F would be ______

5 What **MIL** specifications describe the template shown above? ______ What is an equivalent code specification? ______

6 The **time-delay** symbol of Appendix F describes duration of delay in **milliseconds**, while the **single-shot function** describes **output duration** in ______

7 The aspect ratio (page 131) of the **general logic** symbol is ______

TITLE LOGIC SYMBOL FAMILIARIZATION					DWG. NO. LOG-I	
NAME	DATE	COURSE	GRADE	SCALE 3/4 SIZE	SHEET I OF 2	PAGE 69

With MIL-STD-806 template, fill in the spaces below. Three symbols and titles per line.

SHIFT REGISTER

AND

INCLUSIVE OR

SCHMITT TRIGGER

EXCLUSIVE OR

FLIP-FLOP

SINGLE SHOT

TIME DELAY

AMPLIFIER

TITLE LOGIC SYMBOLS – TEMPLATE FAMILIARIZATION					DWG. NO. LOG-I	
NAME	DATE	COURSE	GRADE	SCALE 3/4 SIZE	SHEET 2 OF 2	PAGE 70

With MIL-STD-806 template, complete below the started logic diagram in Appendix F, page 132.

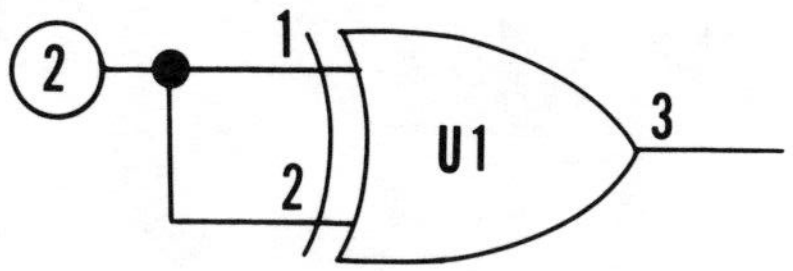

2 1 2 U1 3

TITLE LOGIC DIAGRAM					DWG. NO. LOG-2	
NAME	DATE	COURSE	GRADE	SCALE 3/4 SIZE	SHEET 1 OF 1	PAGE 71

INTRODUCTION TO INTEGRATED CIRCUITS (I.C.)

Integrated circuits are combinations of active elements (such as diodes and transistors) and passive elements (such as capacitors and resistors) functioning as a single item. They generally are made as very small wafers, referred to as "chips" or "dies" 1/16 inch square or even smaller. Within this tiny area, an integrated circuit may contain the equivalent of 50 components or more. Two or more of these circuits are commonly encapsulated in flat rectangular packages known as "flat-packs."

At present there are two principal types of I.C. packages (or chips); (a) the 14- or 16-pin "**dual-in-line**" as shown on page 133 of Appendix G, Fig. 2 (also called the "bug") and (b) the 10- or 14-pin "flat pack" shown on page 134 of Appendix H, Fig. 1 and 2.

The dual-in-line has become more commonly used than the flat pack and requires plated-thru holes (see page 133, Fig. 1). The flat pack is mounted with its leads above the board (see page 134, Fig. 3 through Fig. 3a). It is recommended that all "feed thru" connections on all I.C. boards (both flat pack and dual-in-line) use plated-thru holes rather than buss wire or eyelets.

Both the dual-in-line and flat pack chip utilize a corner **black dot** for orientation purposes; these are illustrated on pages 133 and 134. In particular the dual-in-line dot is located nearest pin 1 and the other pins are ascertained by counting **counter-clockwise** from pin 1.

Appendix G, Fig. 3 shows the internal logic of the various chips, the components of which are known as elements of the I.C.

Complete the exercise at right and continue with page 73. In the lessons that follow, further explanations will be made in greater detail regarding the use of the flat pack and dual-in-line with their requirements for drafting and design applications.

Exercises (Fill in the blanks)

1 Integrated circuits are generally made into small wafers referred to as ____________ or ____________ .

2 The 10-pin I.C. package is known as a ____________ while the 16-pin I.C. package is known as a ____________ .

3 What kind of "feed-thru" connections should a designer select for I.C. boards? ____________ .

4 The black dot on the dual-in-line package lies nearest pin ______ .

5 In locating pin 10, the user should start with the black dot and count in a ____________ ____________ direction.

6 The flat pack comes in either ____________ pins or ____________ pins while the dual-in-line comes in either ____________ pins or ____________ pins.

7 The "bug" is another name for the ____________ package.

8 The package having leads mounted above the board is called the ____________ while the package requiring feed-thru holes is called the ____________ .

TITLE INTRODUCTION TO INTEGRATED CIRCUITS (I.C.)					DWG. NO. ICI–1	
NAME	DATE	COURSE	GRADE	SCALE	SHEET 1 OF 1	PAGE 72

14-Pin Dual-in-Line

Fig. 2 of Appendix G (page 133) illustrates the 14-pin dual-in-line I.C. Notice that the leads are bent downward and in line for easy assembly onto P.C. boards. The dual-in-line leads are inserted into plated-thru holes.

Exercise: Draw three views (front, top, and side) of the 14-pin dual-in-line 2 X size on the right (see page 133). Show all the dimensions, the black dot, and number all the leads as started in Fig. 2. After completing the drawing, fill in the blanks below.

1 The length of the dual-in-line is ________ inches.

2 The width of the dual-in-line is ________ inches.

3 The diameter of the dual-in-line leads is ________ inches.

4 The black dot is located near pin ______ .

TITLE 14–PIN DUAL–IN–LINE PACKAGE INTEGRATED CIRCUIT (I.C.) OUTLINE					DWG. NO. IC–2	
NAME	DATE	COURSE	GRADE	SCALE 2 X SIZE	SHEET 1 OF 1	PAGE 73

Logic diagrams, such as the one drawn on page 71, are made up of several integrated circuit symbols, designated by the letter "U."

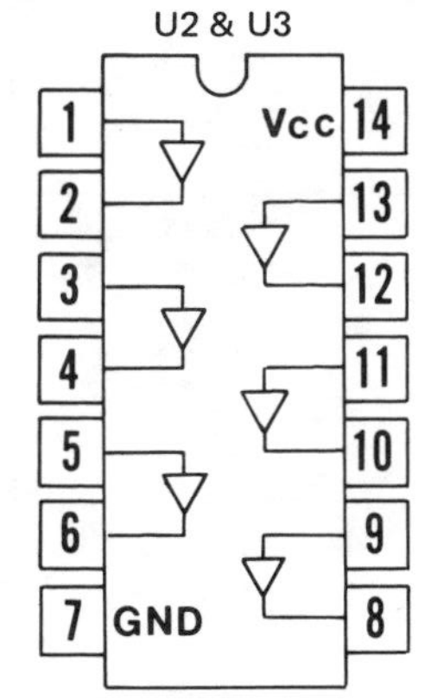

BUFFER
(B74311)

Each "U" symbol represents some "functional elements" **inside** a chip or integrated circuit (I.C.). The chip shown at left is made up of six elements.

The element shown —▷— is called a "buffer."

From page 71 we can count eleven buffers, six in U3 and five in U2. From the chip shown on left we can count six buffers each for U2 and U3 totaling twelve buffers. This implies that Diagram (page 71) is allowing one buffer to remain unused – spare elements are quite common. An example of unused elements can be seen by noting that on page 133, the I.C. called "flip-flop" (F74921) is made up of two identical elements: if only one element is used, as on page 71, it will be denoted as "1/2 of U5."

Each I.C. chip has a ground pin (GND = pin 7) and a voltage pin (Vcc = pin 14) as shown above; the remaining pins are represented on diagrams for example as:

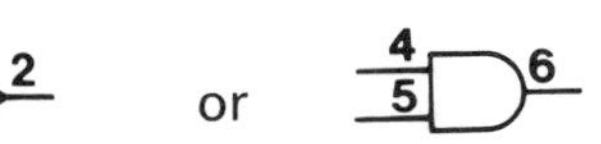

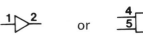

Logic diagrams are drawn from left (input) to right (output). Page 71 has encircled numbers ② on the left and right sides representing **board terminals** or **P.C. tabs** – not chip pins.

Chips are purchased by part numbers such as B74311 or F74921.

Exercise: Use the diagram on page 71 and the I.C. chips on page 133 and complete the following:

1. What letter is used on logic diagrams to represent an I.C.?_______

2. From page 71, how many element symbols are used in U1?______

3. How many functional elements are in the quad gate, Q74513 I.C. chip? ____________________

4. What fraction of the chip U5 is being used?____________

5. On all 14-pin chips, pin 7 is used for ____________.

6. On all 14-pin chips, pin 14 is used for ____________.

7. On page 71, pin 7 of all I.C. are connected to terminal number ________.

8. Chip 1/2 of U5, pin 6 is connected to terminal number________.

9. U3 pin 10 is connected to U4 pin ____________.

10. How many functional elements are in the flip-flop chip, F74921? ____________

11. Complete the drawing on the right of the buffer chip (B74311). Show all elements and pin numbers.

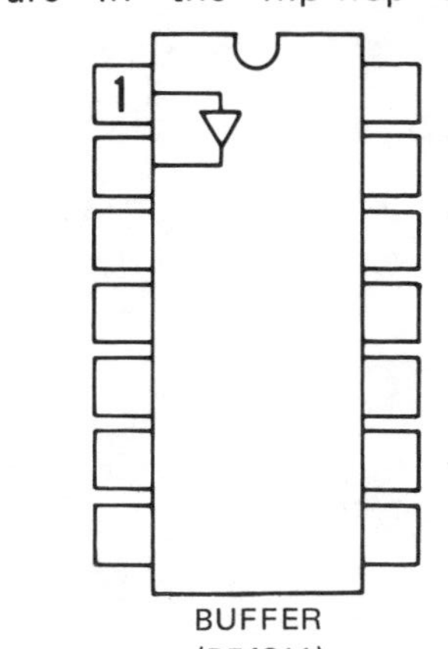

BUFFER
(B74311)

TITLE FUNCTIONAL ELEMENTS FAMILIARIZATION					DWG. NO. IC–3	
NAME	DATE	COURSE	GRADE	SCALE	SHEET 1 OF 1	PAGE 74

Integrated Circuit (I.C.) Design Layout Requirements

To layout an I.C. board, the designer needs the following items: (1) a logic diagram; (2) a board outline; and (3) .100 quadrille grid mylar or graph paper.

I.C. boards should be drawn 2 × size accurately on .100 quadrille grid mylar or graph paper as shown in Fig. A. The layout is viewed from top component side (see page 133, Fig. 1 for reference) showing all the I.C. dual-in-line pads (1/8 O.D. – 2 × size) without components. Circuitry conductors are shown in the usual way:

Component (top) side = solid lines (————)

Circuit (bottom) side = dashed lines (– – – – – – – –)

Fig. A illustrates how long and short conductor lines are drawn. Short conductor lines are used to connect long lines to pads, plated-thru holes or P.C. tabs. It is recommended that long conductor lines be drawn as follows:

Top solid long lines – horizontally (Except for GND and Vcc lines – see page 76A)

Bottom dashed long lines – vertically

Plated-thru holes are used when it is necessary to connect a top conductor line to a bottom conductor line. Plated-thru holes, other than I.C. pads, are indicated thus:

◉ (1/8 O.D. = 2 × size)

In making the layout of the I.C. board, **do not** draw the outline of the I.C. chip – it is not needed.

The hole pattern of I.C. pads and plated-thru holes are located on **grid intersections** as shown in Fig. A and Fig. B.

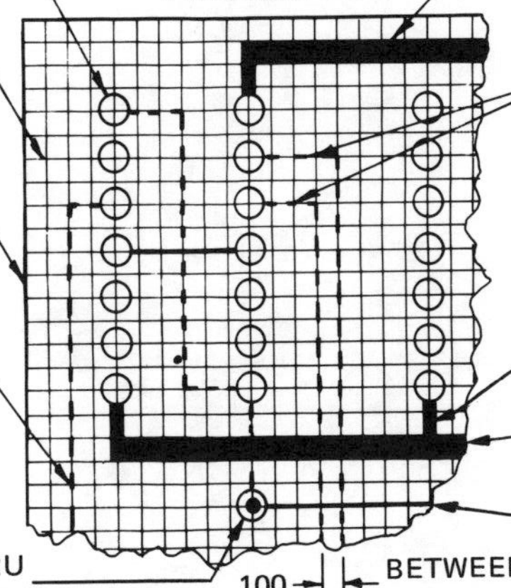

The conductor line widths need not be drawn to scale, however the line widths for ground (GND) and voltage (Vcc) should be heavy – .100 inch (2 × size).

The short conductor line connections to I.C. pin 7 and 14 pads may be drawn narrower than .100 inch; see Fig. A.

Exercise:

1 The three major items needed to lay out an I.C. board are:

a. ______________

b. ______________

c. ______________

2 The quadrille grid lines on mylar or graph paper are spaced ________ apart.

3 What symbol is used to denote a plated-thru hole connecting top and bottom conductors of the board on a layout? ________

4 Line width for GND and Vcc (2 × size) is ________.

5 The hole pattern of I.C. pads and plated-thru holes are located on ________.

6 In making the layout of the I.C. board, (do/do not) ________ draw the outline of the I.C. chip.

7 The I.C. pad diameter is ________ (2 × size).

8 Circuitry conductors shown on component (top) side = ________ lines (________); on the circuit (bottom) side = ________ lines (________).

TITLE INTEGRATED CIRCUIT (I.C.) DESIGN LAYOUT REQUIREMENTS					DWG. NO. IC–4	
NAME	DATE	COURSE	GRADE	SCALE 2 X SIZE	SHEET 1 OF 1	PAGE 75

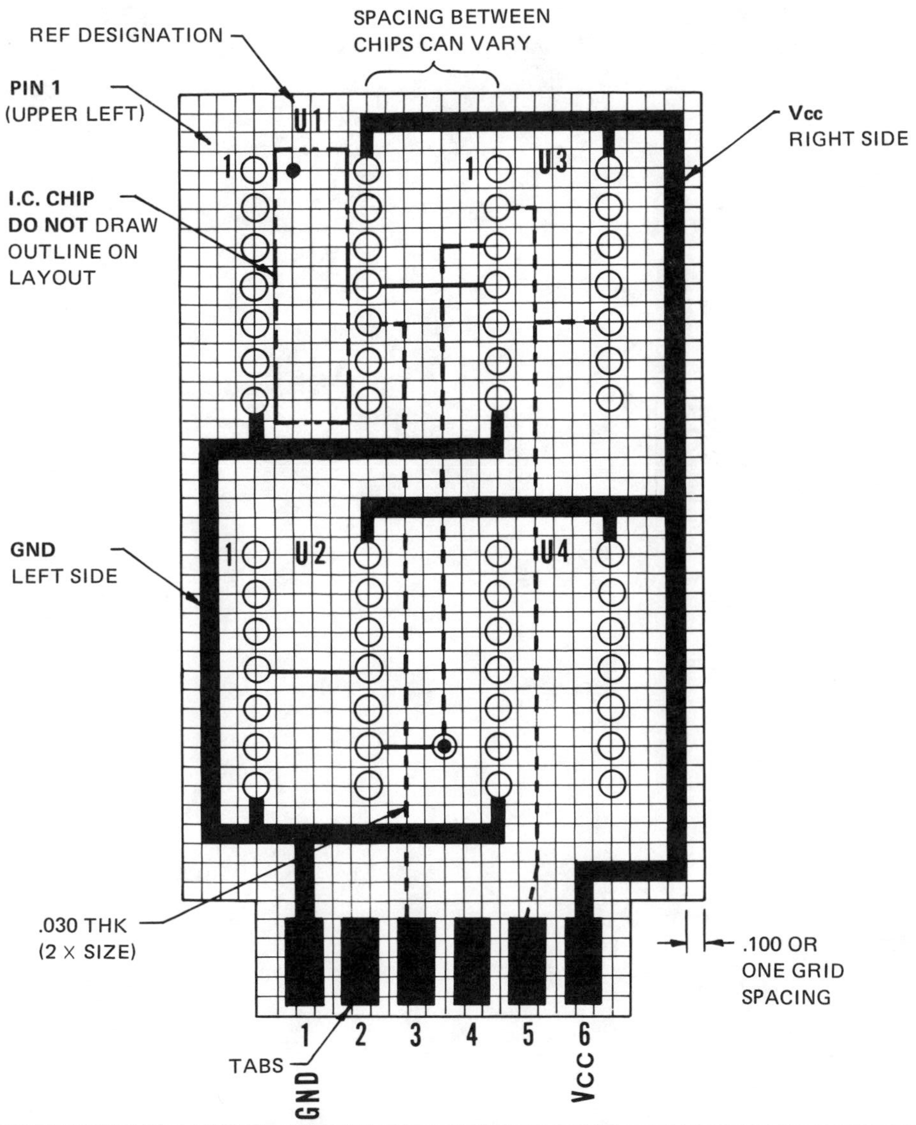

DUAL-IN-LINE LAYOUT TECHNIQUE

To prepare a layout, using dual-in-line chips, proceed as follows, using the illustration at left for reference.

Start by placing all I.C. chips in a vertical position (see phantom view – but do **not** draw chip outline) parallel with the tabs and with the pin 1 of each chip located at the left corner. The spacing between chips can vary depending on the number of chips being used. This method of layout gives the board a uniform, neat appearance and lends itself to ease of assembly in production.

Pin 1 pad is the only pad of each chip that needs to be labeled; then locate the 14 pads of each chip. Label the position of each I.C. chip with a "U" number such as "U1," "U2," "U3," etc. The rough layout can be filled in now by showing all the conductors and plated-thru holes using the logic diagram as a reference.

To complete the "finishing touches", do the ground (GND) and voltage (Vcc) conductor lines first. These lines are placed on top of the board with the GND lines proceeding from the left GND tab and the Vcc lines proceeding from the right Vcc tab as shown. Leave one grid edge spacing around the board outline and between conductor lines (.100).

Both solid and dashed lines should follow the grid pattern and all bends should be made with 90 degree corners. Plated-thru holes should **not** be located under an I.C. – they should be visible for inspection purposes from both sides of the board.

Exercise: Complete the following:

1. I.C. chips should be placed in a ____________ position, parallel with the tabs.
2. The I.C. Chip is oriented so that the pin 1 is located in the ____________ corner.
3. The GND conductor originates from a tab located on the ____________ side of the board.
4. The Vcc conductor originates from a tab located on the ____________ side of the board.
5. Conductor lines should be bent with ____________ degree corners.

TITLE DUAL-IN-LINE LAYOUT TECHNIQUE					DWG. NO. IC-5	
NAME	DATE	COURSE	GRADE	SCALE 2 X SIZE	SHEET 1 OF 2	PAGE 76A

Exercise:

Study the logic diagram below, and complete the started layout at right by filling in the missing conductor lines.

First. Connect all GND conductors on the pads for pins 7 of all the I.C. chips to tab 1 in the same manner as was shown in the previous exercise (page 76A).

Second. Follow the same procedure of page 76A to connect all the Vcc conductors on the pads for pins 14 of all the I.C. chips to tab 6.

The last step is to connect **four missing** conductor lines as per the logic diagram; it probably is simpler to proceed from left to right. (Hint: one of the four missing lines will require a plated-thru hole).

4 DUAL-IN-LINE LAYOUT
(2 × SIZE)

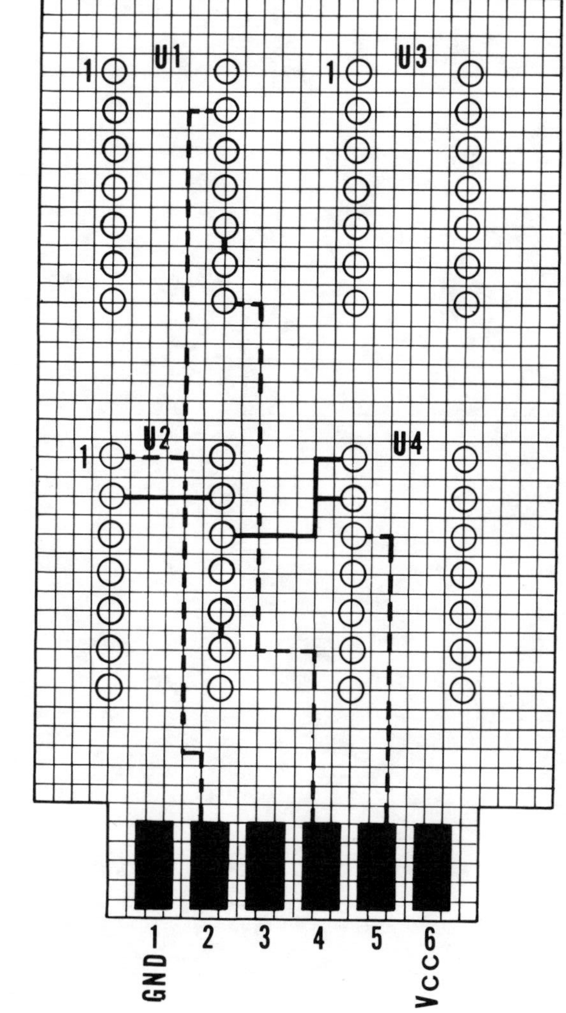

LOGIC DIAGRAM

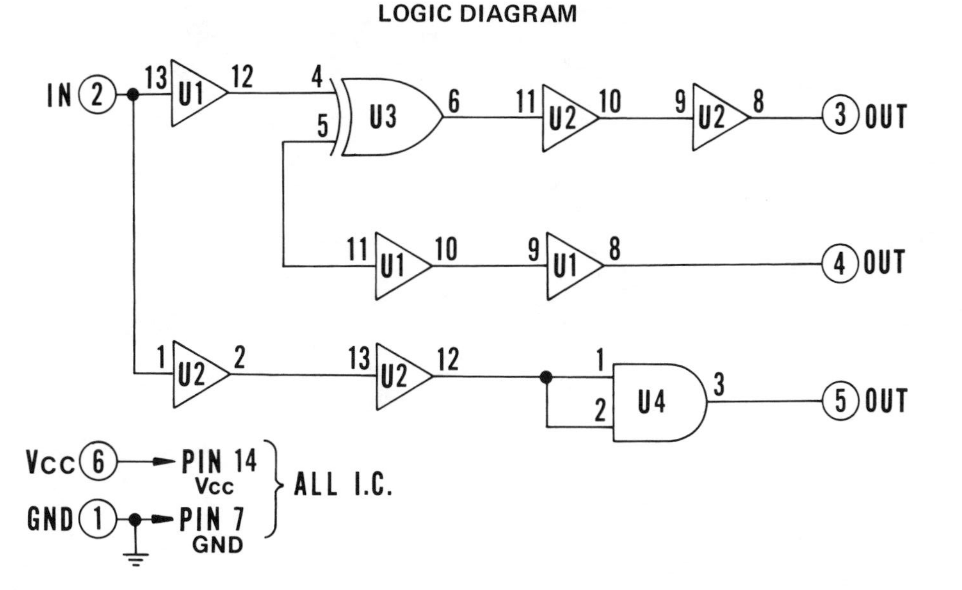

TITLE DUEL-IN-LINE LAYOUT TECHNIQUE					DWG. NO. IC-5	
NAME	DATE	COURSE	GRADE	SCALE 2X SIZE	SHEET 2 OF 2	PAGE 76B

Exercise:

Complete the I.C. board design layout started at right. To complete the layout, use the logic diagram drawn on page 71 and the I.C. chips of page 133 (Fig. 3).

Notice that there are five 14 pin dual-in-line chips mounted on the board (U1 through U5).

The board is drawn 2 × size on a .100 Grid.

The tabs are numbered from 1 through 14, corresponding to the logic diagram inputs and outputs. The same numbered tabs are both on top and bottom of the board and conductors can extend from either or both sides of the tab.

Recall that on page 76B you started your layout by drawing in the GND and Vcc conductor lines first (on top); this procedure should be followed again in the layout on your right.

Use the logic diagram (page 71) as a reference to complete the layout, showing all solid and dashed lines plus all plated-thru holes as required.

HIGH DENSITY CONDUCTOR LAYOUT FOR I.C.s
Another method used for routing conductors in a high-density I.C. board layout is to run the conductor in between the pads as shown on the left. (The pads are flatted or sometimes the conductor path is necked down.)

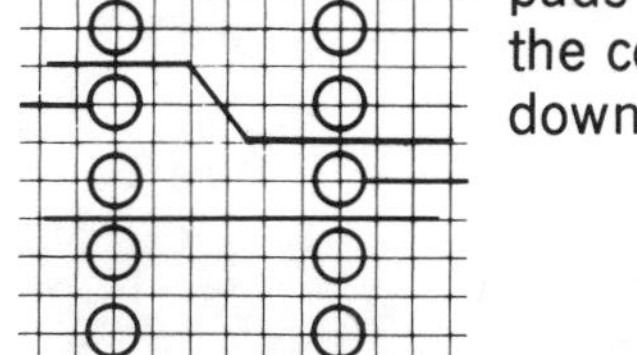

U1 U3 U4

U2 U5

1 2 3 4 5 6 7 8 9 10 11 12 13 14

GND Vcc

TOP ——— BOTTOM — — — PLATED-THRU HOLE

TITLE I.C. BOARD DESIGN LAYOUT — 5 DUAL-IN-LINE					DWG. NO. IC—6	
NAME	DATE	COURSE	GRADE	SCALE 2 X SIZE	SHEET 1 OF 1	PAGE 77

The 14-Pin Flat-Package I.C. Fig. 2 of Appendix H (page 134) illustrates the 14-pin flat-package. The flat-package is mounted with its leads above the board (see page 134, Fig. 3).

Exercise: Draw the front and top views of the 14-pin flat-package (page 134, Fig. 2) 10 x size below. Label the pins 1, 2, 3, etc., as shown with the black dot between numbers 1 & 2.

Answer the following. The 14-pin flat-package has a:

1 leads length of________ inches.

2 leads width of ________inches.

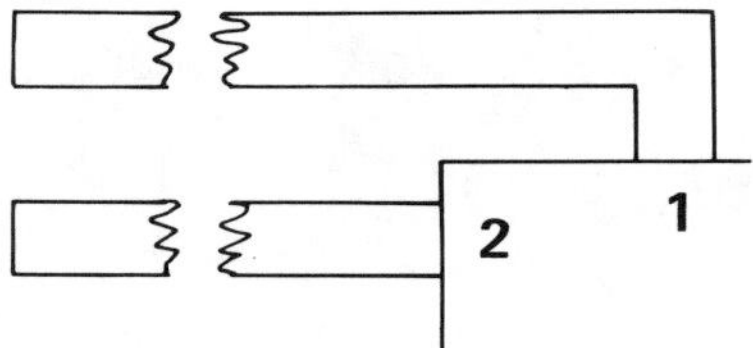

TITLE 14–PIN FLAT–PACKAGE INTEGRATED CIRCUIT (I.C.) OUTLINE

DWG. NO. IC–7

PAGE 78

NAME	DATE	COURSE	GRADE	SCALE 10 X SIZE	SHEET 1 OF 1

I.C. HOOK-UP, DISTRIBUTION OF LEADS – FLAT PACKAGE

Integrated Circuit (I.C.) packages are often layed out in rows with certain common leads hooked "in parallel." For example, in most 14-pin flat-package the **power lead** is pin **4** (Vcc); this pin (or lead) is usually connected in parallel to pin **4** leads of the adjacent flat-packages. Similarly, all the **ground leads** or pin **10** (GND) are connected in parallel.

In the layout at right, the parallel connection of all the **4** pins has been started. Similarly, all the **10** pins are connected in parallel as are all the **input leads** or pin **3**. Each set of pins is **separately** hooked in parallel. The **4** pins and **10** pins have been laid out with connections on the **top side** of the board; the **3** pin connections are made on the **bottom side** via **plated-thru holes.**

The remaining unused leads are normally soldered to the nearest **blank** (unconnected) **lap joint** or **land** (see Appendix H, Fig. 3.).

Exercise 1:

a. Which number is used to represent the **power lead** (Vcc)?_____

b. Which number is used to represent the **ground lead** (GND)?_____

c. Which number is used to represent the **input lead?**_____

Exercise 2: Connect the flat-packages A1, A2, & A3 below so that the pins **3, 10, & 4** are each separately hooked in parallel. Draw in all leads & cross-hatching.

Key: Top side conductors: [hatched]

Bottom (hidden) side conductors: [dashed]

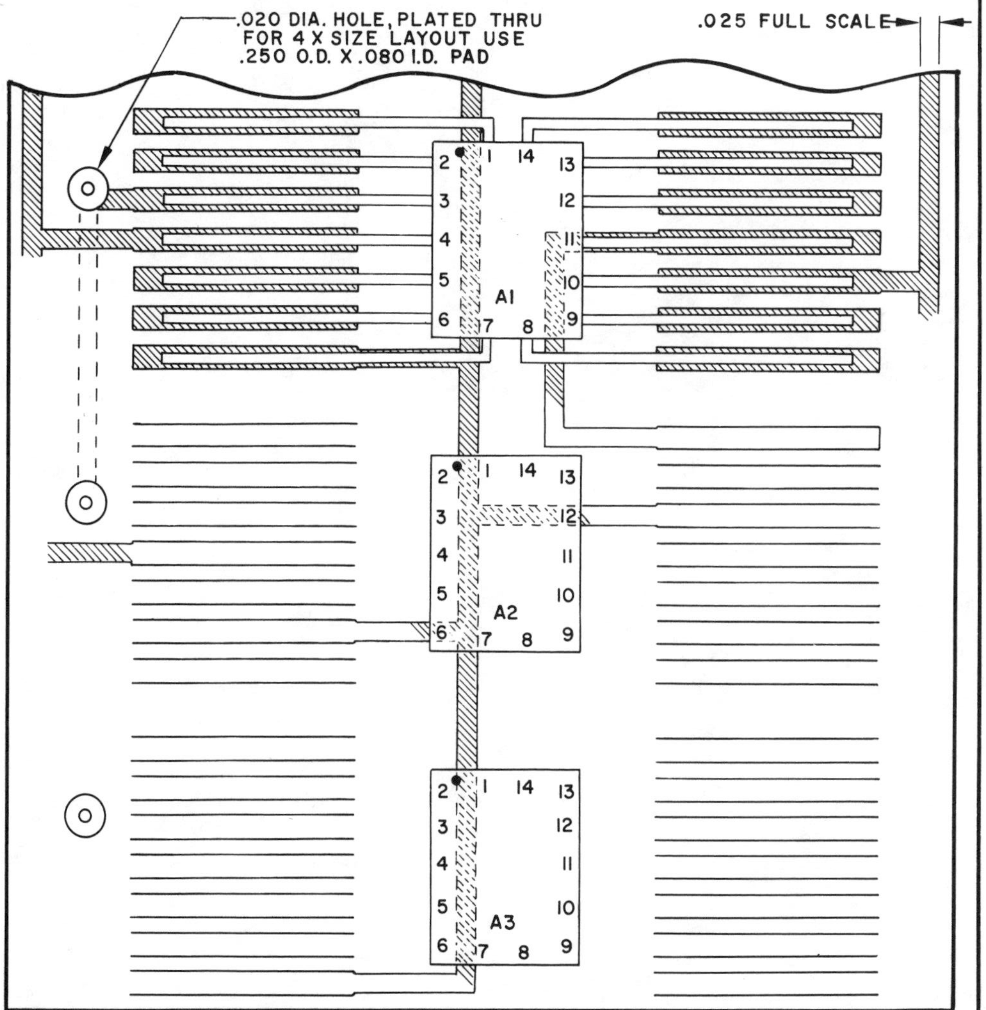

TITLE INTEGRATED CIRCUIT HOOK-UP (FLAT PACKAGE)					DWG. NO. IC-8	
NAME	DATE	COURSE	GRADE	SCALE 4X SIZE	SHEET 1 OF 1	PAGE 79

INTRODUCTION

The MIL-STD-1313A definition of a microcircuit is: "A small circuit having a high equivalent circuit element density, which is considered as a single part composed of interconnected elements on or within a single substrate to perform an electronic circuit function."

Before one can prepare a layout of a hybrid microcircuit, some basic concepts need to be covered. These are: Substrates, thin and thick film, chip components, adhesives, wire bonding, and vias.

SUBSTRATES (MIL-STD-1313A)

"The supporting material upon or within which the elements of a microcircuit are fabricated or attached" is called a substrate.

An example of a **hybrid microcircuit** is shown in Fig. 1, which is an enlarged portion (approx. 6 × size) of a pacemaker unit. The LSI (large-scale integration) in this unit has 56 bonding pin wires and is actually only .160 × .250 inches. The transistor of Fig. 1 is actually only .050 inch square. Since hybrid microcircuits are so small, the layouts are prepared at a 20 to 1 scale.

THIN AND THICK FILM

There are basically two types of hybrid microcircuits being used—the **thin film** and the **thick film** circuit. Both circuits use the same layout technique.

THICK FILM CIRCUIT

A ceramic substrate, bearing circuit traces obtained by screen printing and firing, specially prepared thick film pastes, usually made of platinum and/or gold, is called a thick film circuit.

Three basic pastes are employed:
1. Conductive 2. Dielectric 3. Resistive

That is, thick film is used to form conductors, resistors, and capacitors.

THIN FILM CIRCUIT

A ceramic substrate, bearing circuit traces, obtained by plated or evaporated gold conductive patterns is called a thin film circuit.

THIN FILM RESISTORS

This film can also be used to form thin film resistors directly on a substrate circuit and then be trimmed to the required tolerance by laser beams.

Hybrid microcircuits operate at very low power. For example, the pacemaker unit (Fig. 1) operates at 2.8V, with a very small battery that will last for almost 10 years.

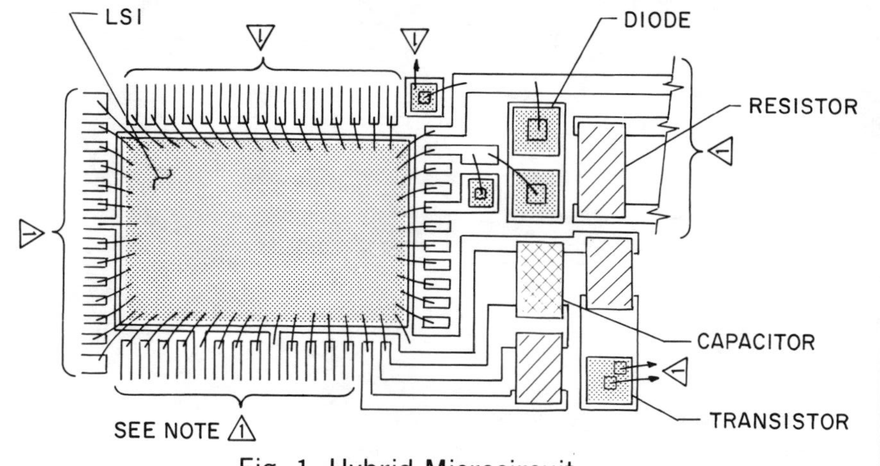

Fig. 1 Hybrid Microcircuit (Top view)

Portion of an enlarged hybrid microcircuit, from a pacemaker unit (approx. 6 × size). Courtesy Pacesetter Systems Inc.

Note △1 Interconnects to remainder of pacer electronics circuitry.

Exercise

1. What is a substrate? ______________________________

2. What is a thick film circuit? ______________________________

3. What are the three pastes that are employed?
   1. ______________________________

4. What is a thin film circuit? ______________________________

TITLE HYBRID MICROCIRCUITS					DWG. NO. HMC-1	
NAME	DATE	COURSE	GRADE	SCALE	SHEET 1 OF 4	PAGE 80

CHIP COMPONENTS

It is good practice, when preparing a hybrid microcircuit layout, to represent symbolically the chip components by rectangular shapes which roughly correspond to the actual shape of the components as shown below.

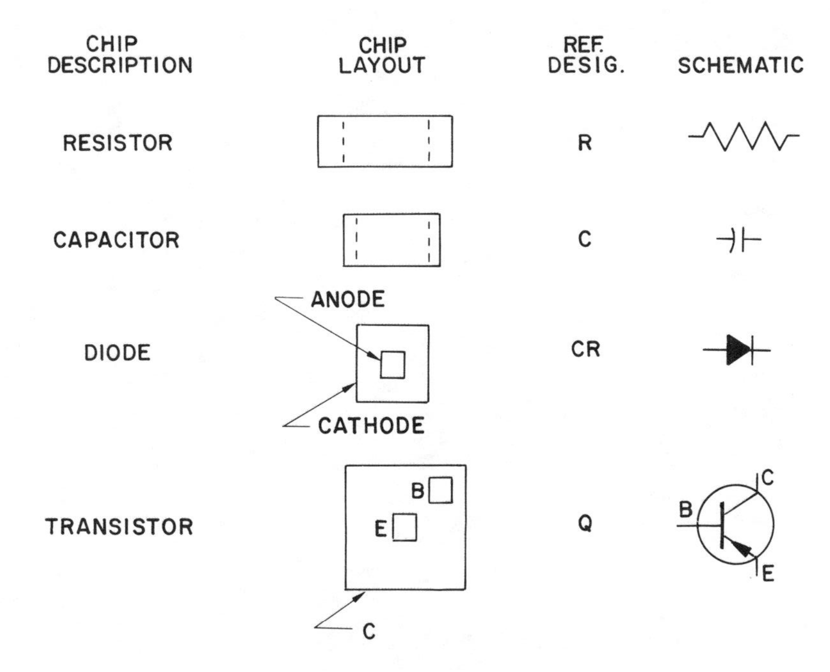

ADHESIVES

Conductive and nonconductive adhesives are used in the assembly of hybrid microcircuits to provide both electrical and mechanical bond.

In the pacemaker unit (Fig. 1 of page 80), all of the chip components were bonded to the conductors with conductive epoxy.

The emitter (E) and the base (B) of the Transistor in the pacemaker unit have **bonding wires** which are connected to other circuitry in the unit (not shown) while the collector (C) which is the entire square of the transistor chip is attached to the circuit with *conductive epoxy* to provide electrical bond. (See enlarged view, Fig. 1.)

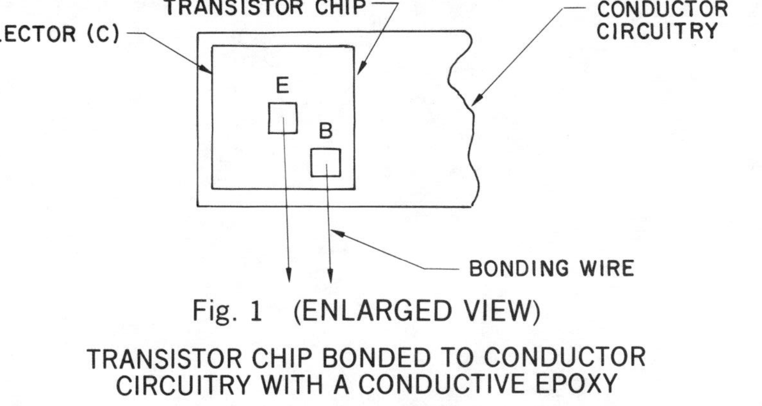

Fig. 1 (ENLARGED VIEW)

TRANSISTOR CHIP BONDED TO CONDUCTOR CIRCUITRY WITH A CONDUCTIVE EPOXY

A nonconductive epoxy-type adhesive is used when electrical conductivity is not required. An example would be attachment of passive component chips which will subsequently be electrically interconnected by **wire bonding**. See Fig. 2.

RESISTOR CHIP
NONCONDUCTIVE EPOXY
CONDUCTOR
BONDING WIRE
SUBSTRATE

Fig. 2

RESISTOR CHIP BONDED TO CONDUCTOR WITH A NONCONDUCTIVE EPOXY

WIRE BONDING

Gold and aluminum wire are used for electrical interconnection and jumpers in the assembly of hybrid microcircuits.

All of the wire bonding, used in the pacemaker unit, is 1.0 mil aluminum.

Exercise:

1. What shape do chip components roughly correspond to in hybrid microcircuit layouts? ______
2. What kind of adhesive is used when a component chip requires a good electrical bond? ______

TITLE HYBRID MICROCIRCUITS					DWG. NO. HMC-I	
NAME	DATE	COURSE	GRADE	SCALE	SHEET 2 OF 4	PAGE 81

PREPARATION OF HYBRID MICROCIRCUIT LAYOUT

The layout on the right (Fig. 2) is an enlarged portion of the final layout of the hybrid microcircuit (pacer unit — Fig. 1, page 80). The layout is prepared on 0.1 inch grid at 20 to 1 scale, and checked against the original schematic.

Since this particular unit was designed on several planes called multilayer boards having a large density of conductors, VIAs were used.

VIAS (See sketch, Fig. 1 below)
When multilayer microcircuits are used they are separated by a dielectric layer (insulater). The electrical interconnections between conductor levels are accomplished by small through-holes, similar to plated through-holes in printed circuit boards, or windows called "VIAS" (pronounced "VEEAHS"). The vias are indicated on the layout with a small ▲.

Components mounting (END-BANDS)
Always check the recommended mounting areas of the resistors and capacitors, also called END-BANDS. For example, in Fig. 2, the resistors' End-Band is 0.015 inch, and the capacitor's is 0.010 inch. The edge distance between all components and conductor pads is 0.005 inch, as shown.

Fig. 1. Hybrid multilayer boards

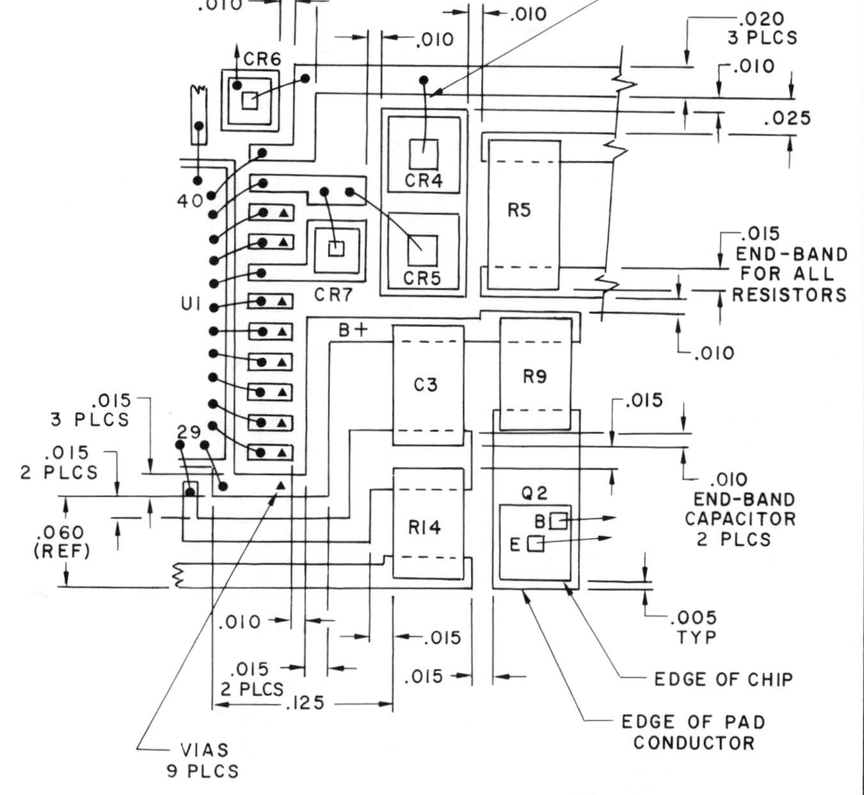

Fig. 2. Hybrid microcircuit layout

The above layout dimensions are for use in solving exercises (page 83). Normally these dimensions would be omitted.

Exercise.
1. What scale is used in preparing a hybrid layout? ______
2. What are vias? ______
3. What are end-bands? ______

TITLE HYBRID MICROCIRCUITS					DWG. NO. HMC-1	
NAME	DATE	COURSE	GRADE	SCALE	SHEET 3 OF 4	PAGE 82

Exercise.

Use Fig. 2, page 82, for reference to duplicate and complete the grid layout on your right (page 83). Show all pad connections and draw patterns with chip components at scale 20 to 1.

Start the layout by locating R14* first. The dimensions of the chips are as follows:

CR4, CR5, and Q2 = 0.050 inch square.

C3 = length 0.080
width 0.050
end band 0.010

R9 and R14 = length 0.075
width 0.050
end band 0.015

R5 = length 0.100
width 0.050
end band 0.015

The above actual sizes have to be increased 20 times for the layout.

Show E and B of Q2, all wire bonding, vias, and the B+ approximately as shown.

All vias are indicated with a small ▲.

***Hint 1:** Top of R14 is "in line" with first horizontal line below U1 and .125 from the vertical line below U1.

Hint 2: E and B of Q2 are approx. .010 square — actual size.

Hint 3: The anodes for CR4 and CR5 are .020 — actual size.

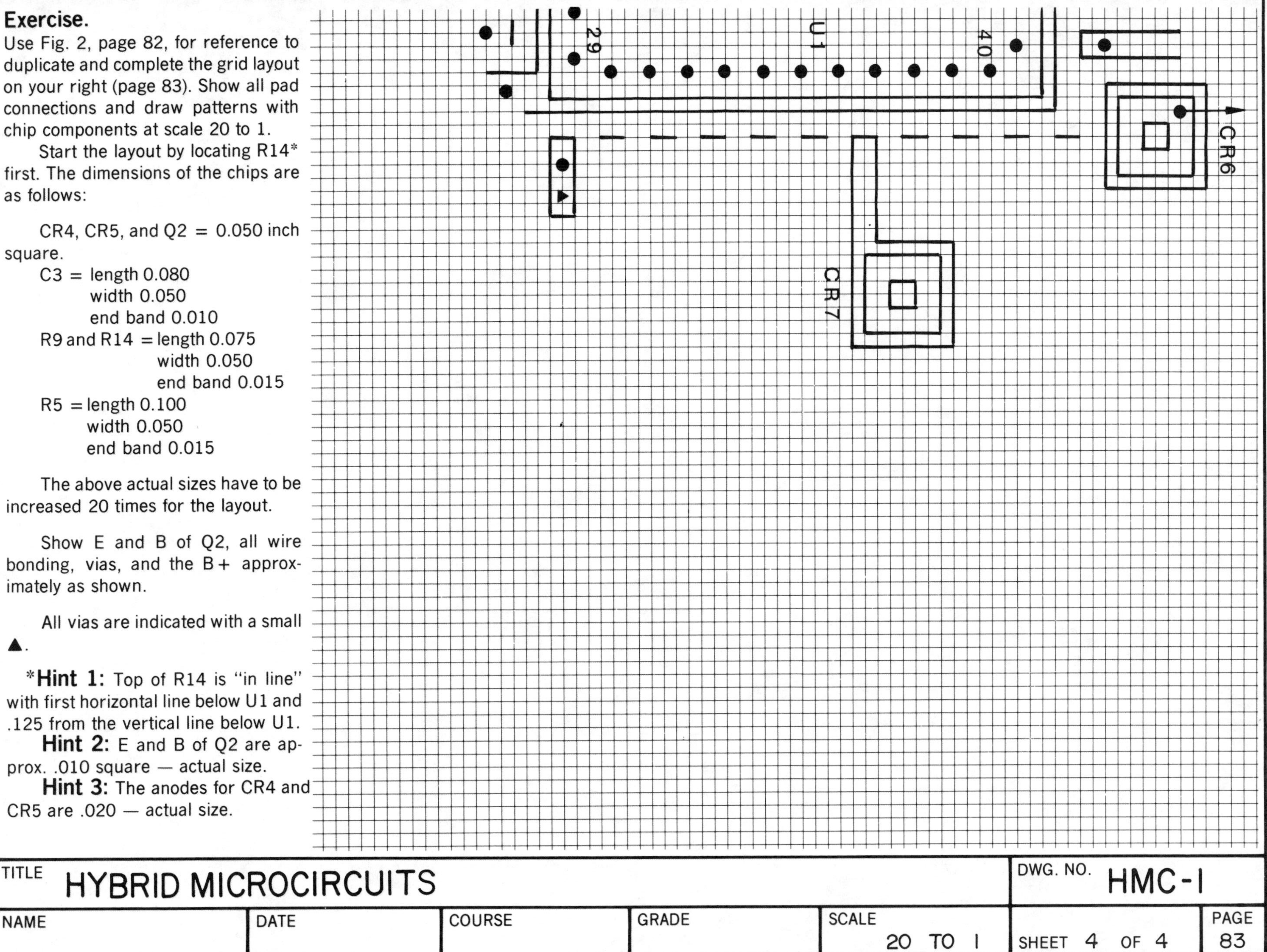

TITLE HYBRID MICROCIRCUITS					DWG. NO. HMC-I	
NAME	DATE	COURSE	GRADE	SCALE 20 TO I	SHEET 4 OF 4	PAGE 83

CAD INTRODUCTION

CAD (Computer-Aided Drafting/Design) is an electronic method of doing drafting or design work, using computer techniques. The drafting principles are the same as in manual drafting, and should be understood **BEFORE** attempting CAD. The lessons which follow will enable you to grasp some of the main concepts of CAD by both generalizing and specifying a particular computer graphics system. (There are over three dozen systems.)

In Fig. 1 below, most of the hardware items shown are a part of all systems. There are three principal CAD types: the **"micro"** CADs, which don't have the joystick or light pen; the **"mini"** CADs, some of which don't have light pens; and the **"mainframes"**, most of which have all items shown. Basically, the CAD operator draws (e.g., a schematic) with a joystick, function, and keyboard while using the puck to select symbols in the menu. Micro systems may use a scriber/stylus as a combination joystick-puck. The scriber/stylus can be used to sketch on the digitizer (see Fig. 1) which appears on the CRT screen, and when ready, a "printout" is made by the plotter.

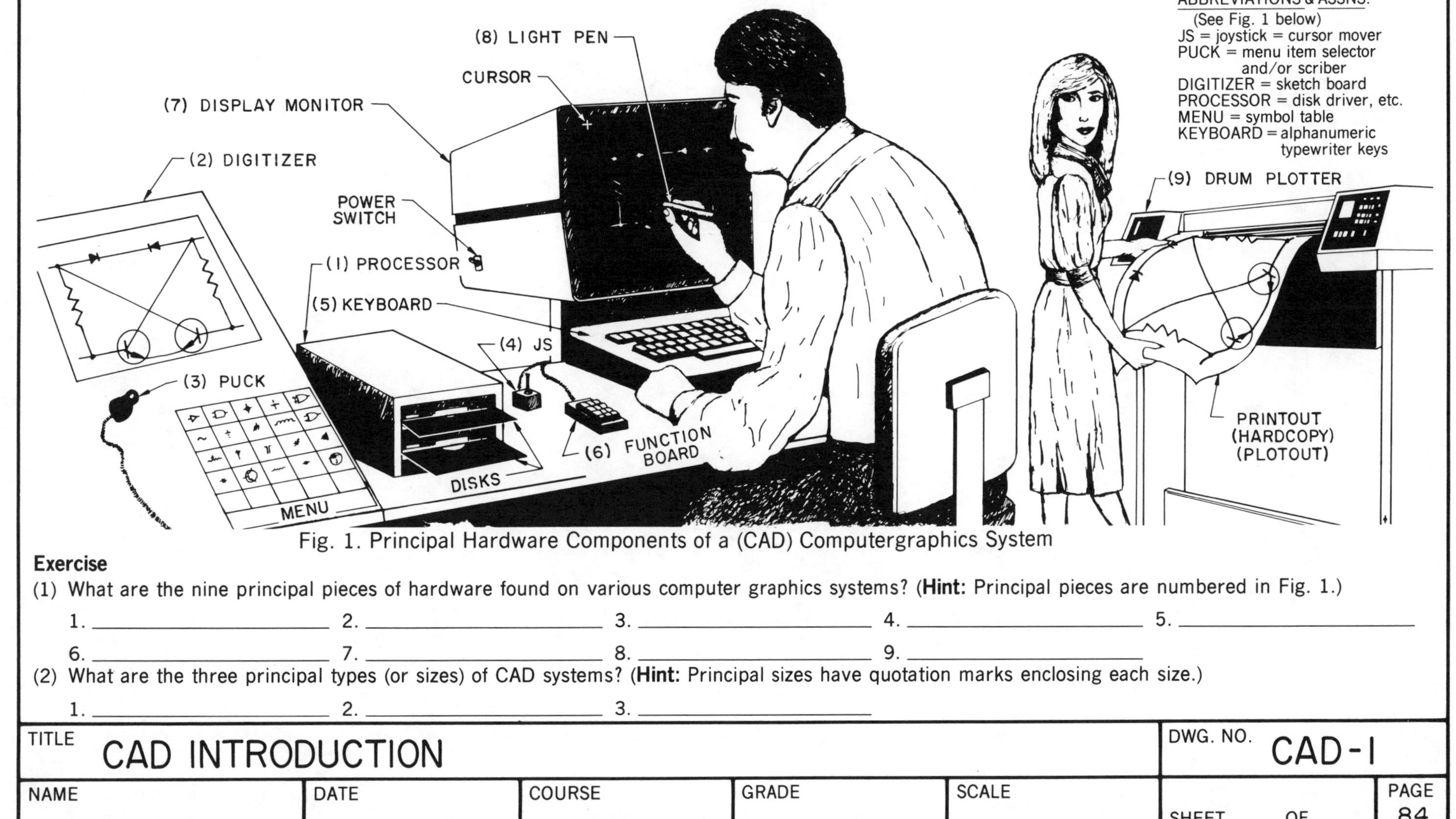

Fig. 1. Principal Hardware Components of a (CAD) Computergraphics System

Exercise

(1) What are the nine principal pieces of hardware found on various computer graphics systems? (**Hint:** Principal pieces are numbered in Fig. 1.)

1. ____________ 2. ____________ 3. ____________ 4. ____________ 5. ____________

6. ____________ 7. ____________ 8. ____________ 9. ____________

(2) What are the three principal types (or sizes) of CAD systems? (**Hint:** Principal sizes have quotation marks enclosing each size.)

1. ____________ 2. ____________ 3. ____________

TITLE CAD INTRODUCTION					DWG. NO. CAD-I	
NAME	DATE	COURSE	GRADE	SCALE	SHEET OF	PAGE 84

CAD HARDWARE (See Fig. 1, CAD-1)

1. **Processor** — Most, but not all, processors are placed behind the keyboard with the disk driver as part of the unit. However, some disk drivers stand alone. Processors can also use tape reels or cassettes instead of disks. The operator may have to change disks/tapes occasionally to gain access to various "memory-library files."
2. **Digitizer** — The average digitizer is about the size of a T-square board and, similarly, is used to sketch drawings free hand which appear straight-lined on the CRT and hard copy. It also includes a symbols/components menu.
3. **Puck** (Scriber/stylus) — Like a drawing pencil, the puck is used to make sketches on the digitizer and to select symbols or components from the menu.
4. **Joystick** — It is like a 4-way toggle switch that moves the cursor (+) on the screen, and together with the function/keyboard it sketches lines on the CRT. (JS = move cursor by joystick.)
5. **Keyboard** — This is like a typewriter with some extra keys. It is used to enter commands to draw, type, store, plot, etc.
6. **Function Board** — An additional keyboard used to "in-put" data or "recall" a function. A menu table is often regarded as part of the function board rather than part of the digitizer.
7. **Display Monitor** — This is also called the CRT, or screen; it is like a TV screen which displays the drawing as it is being sketched.
8. **Light pen** — Used against the screen. The light pen moves the cursor to generate lines, erase, or point to question "prompts" on the screen to indicate answers or decisions.
9. **Plotters** — These usually come as "flat-bed" (small "B" size) or "drum" plotters for "C" size or larger drawings. It plots the final drawing on "hard copy" with ink, using several colors or several different line widths.

Exercise.
Complete the front view of the puck in Fig. 2 at right, from the sketch of Fig. 1. Use full scale; omit dimensions.

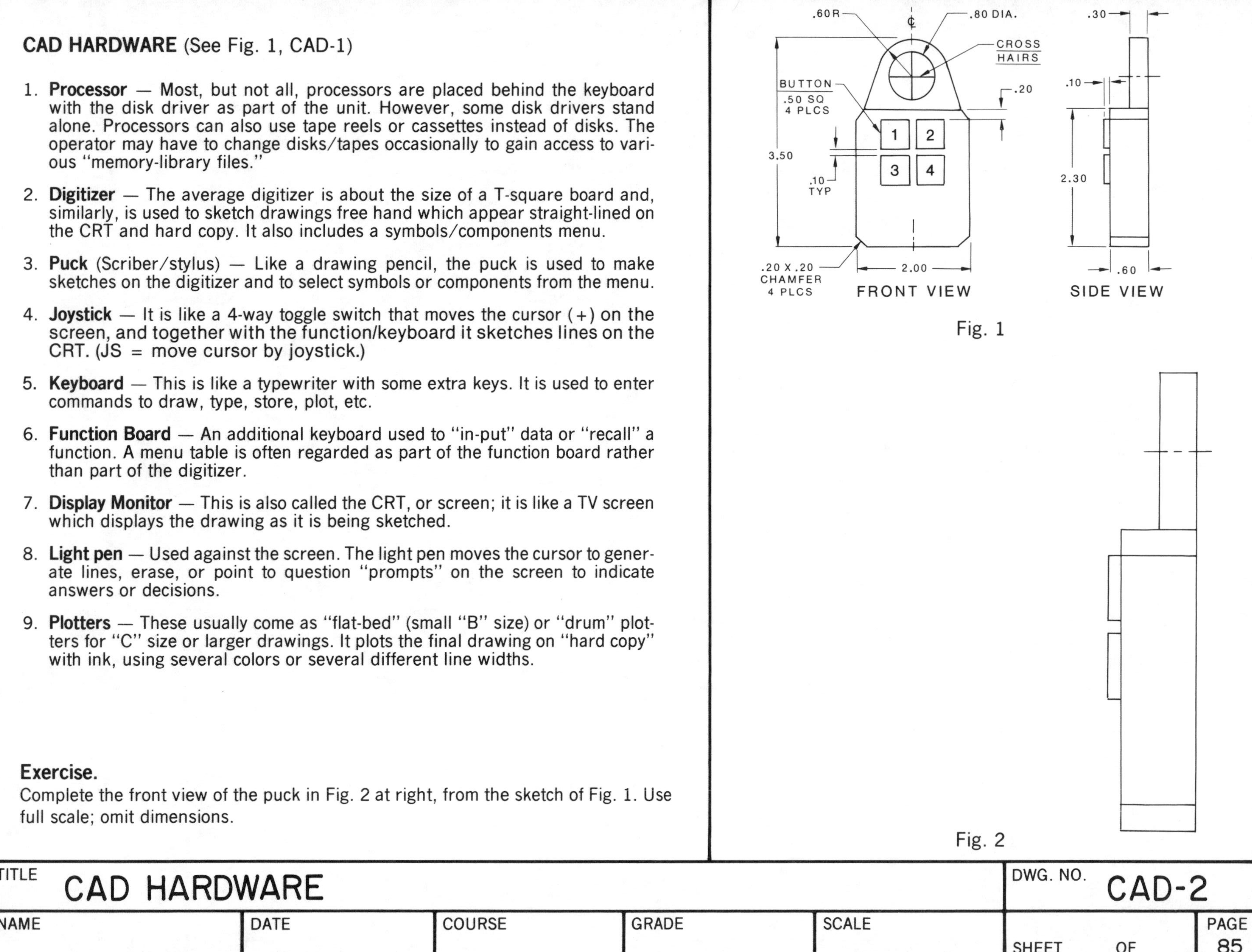

Fig. 1

Fig. 2

TITLE CAD HARDWARE					DWG. NO. CAD-2	
NAME	DATE	COURSE	GRADE	SCALE	SHEET OF	PAGE 85

CAD ADVANTAGES

CAD has come into wide use because, compared to manual drafting, it is less tedious and a more efficient means of producing drawings. There are at least twelve ways in which a drawing is produced more efficiently through CAD:

1. Less time is spent on lettering and line weights.
2. New drawings can, more quickly, be developed from old drawings, which are recalled from the memory-library.
3. Drawing symbols/components repeatedly can be executed instantly by CAD compared to using a template.
4. Erasing large areas is faster by CAD, with no "ghost lines."
5. Deleting specific crisscrossing lines or curves can be achieved without expunging other nearby lines, as when using an eraser manually.
6. Rotating symbols and especially components for PCB optimal placements is easier with CAD.
7. Drawing "to scale" is faster, and *changing* the scale is automatic.
8. One half of symmetrical drawings can be "mirrored" to save drawing time (see exercise at right).
9. Properly spaced crosshatching can be done automatically.
10. Drawing circles/holes/components, evenly spaced on a grid, is also done automatically (known as the "step and repeat" process).
11. Systems with "3D" capability can take a two-view drawing and prepare an auxiliary or third view automatically.
12. Some systems have "automatic routing and/or automatic placing" capabilities. This means that PCBs can "largely" have their components optimally placed for optimal conductor routing—shortest conductors with least crossovers.

To illustrate the "mirroring" technique of CAD, one could use the schematic of page 37 or page 84 and just draw the *left half* by CAD. One would also have to indicate a "centerline-fulcrum" that acts like a hinge around which the left half is revolved to the right side. See Fig. 1 below. After the left half is drawn, a properly coded command automatically draws the right half.

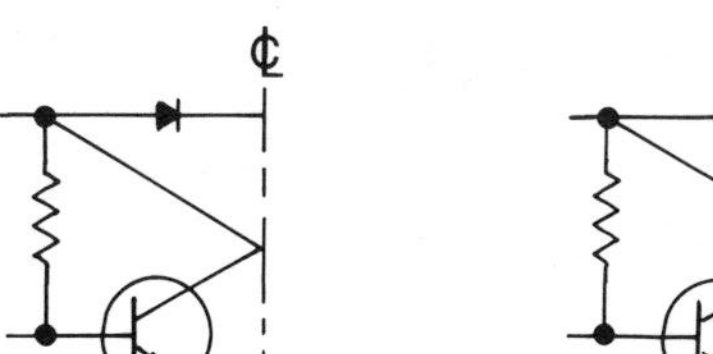

(*a*)

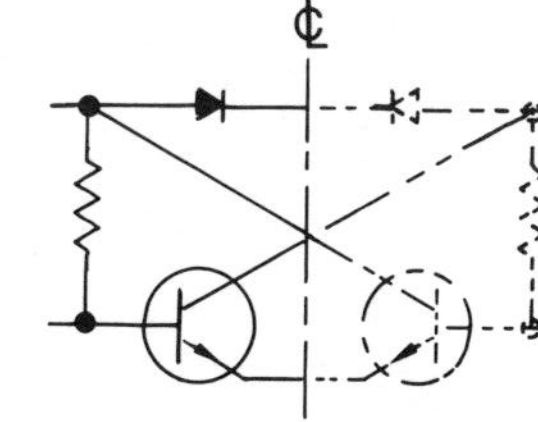

(*b*)

Fig. 1

The right side of Fig. 1(*b*) would look like the mirror image of the left side, using the CAD mirror technique.

Exercise.
Sketch the right side of Fig. 2 in phantom lines to represent the mirror image as in Fig. 1(*b*), which could be done in solid lines automatically by CAD.

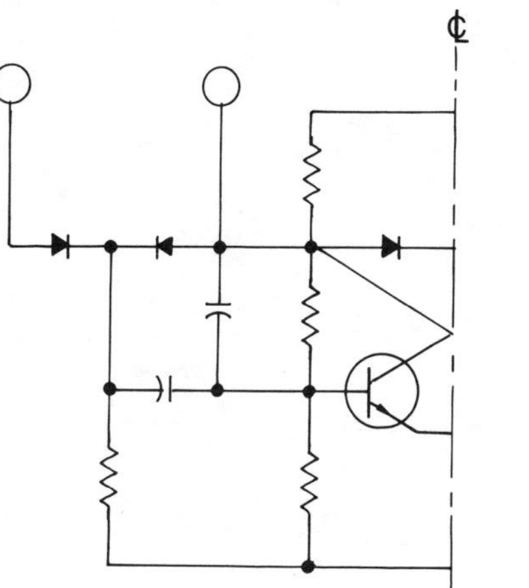

Fig. 2

TITLE CAD ADVANTAGES

DWG. NO. CAD-3

NAME	DATE	COURSE	GRADE	SCALE	SHEET OF	PAGE 86

Exercise. Opposite each of the 21 items listed below, representing either a CAD hardware piece or a CAD advantage, indicate the word or phrase selected from the right side with which it is best associated. (**Hint:** For reference, see CAD 2 and CAD 3.) Letter the entire word or phrase as elegantly as you can.*

1. Processor —
2. Digitizer —
3. Puck —
4. Joystick —
5. Keyboard —
6. Function board —
7. Display monitor —
8. Light pen —
9. Plotter —
10. Lettering and line weight —
11. New drawings from old —
12. Symbol/components drawn repeatedly —
13. Erasing —
14. Deleting —
15. Rotating —
16. Scaling —
17. Mirroring —
18. Crosshatching —
19. Evenly spaced grid items —
20. 3-D systems —
21. Automatic routing and/or placement —

- Screen pointer
- Additional keyboard
- Cursor mover
- Hard copy
- Disk driver
- Scriber/stylus
- Typewriter
- TV screen
- Sketch pad
- Optimal component placement for optimal conductor routing
- Step and repeat
- Half the drawing done automatically
- Revolving for optimal placement
- No ghost lines
- Typing
- Memory-library
- Executed instantly compared to a template
- Expunges specific lines
- Change of scale is automatic
- Lines properly spaced automatically
- Automatic auxiliary views

*CAD lettering, by contrast, is typed to make all drawings appear more uniform and executed more efficiently.

TITLE CAD INTRODUCTION REVIEW					DWG. NO. CAD-4	
NAME	DATE	COURSE	GRADE	SCALE	SHEET OF	PAGE 87

CAD SOFTWARE

In brief, "software" means the programs which have been put on disks (or tapes, cassettes, etc.) that cause the CAD hardware to perform as required. The CAD operator's main concern with software is to be certain that the proper disk is in the processor for the task at hand and to be prepared to change disks if needed. This in turn is related to what is in storage on which disk.

Most CAD operators will work at their terminal (CAD work station) with a number of items — symbols, old drawings, text notes, etc., already in storage on a disk. Symbols, for example, are generally on file, often in what is called a menu (see the two menus in Appendixes I and J). The menu, in small "micro" CAD units, is usually on the digitizer board. Selection of symbols then is made by activating the puck or stylus on a particular menu square symbol, which digitizes the board to recall the symbol to the CRT screen.

Exercise 1

In the started two-view digitizer drawing below, the schematic of page 37/84 has already been drawn in the upper left corner; draw the corresponding component view of the PCB in the upper right corner at full scale. Also, show menus of Appendix I in lower left corner and Appendix J in lower right corner. Draw menus at 1/10 scale; show squares only — no symbols.

Exercise 2

In the started two-view drawing below, complete the front view of a disk whose dimensions are .04 x 5.5 O.D. x 1.5 DIA CTR hole. Show dimensions — scale: 1/5.

Exercise 3

In the started three-view drawing of a processor housing below, capable of accommodating the above disk (2 plcs), draw the top view above the side view. Scale: 1/5. The top view should be sectioned, like the side view, to show the disk in place.

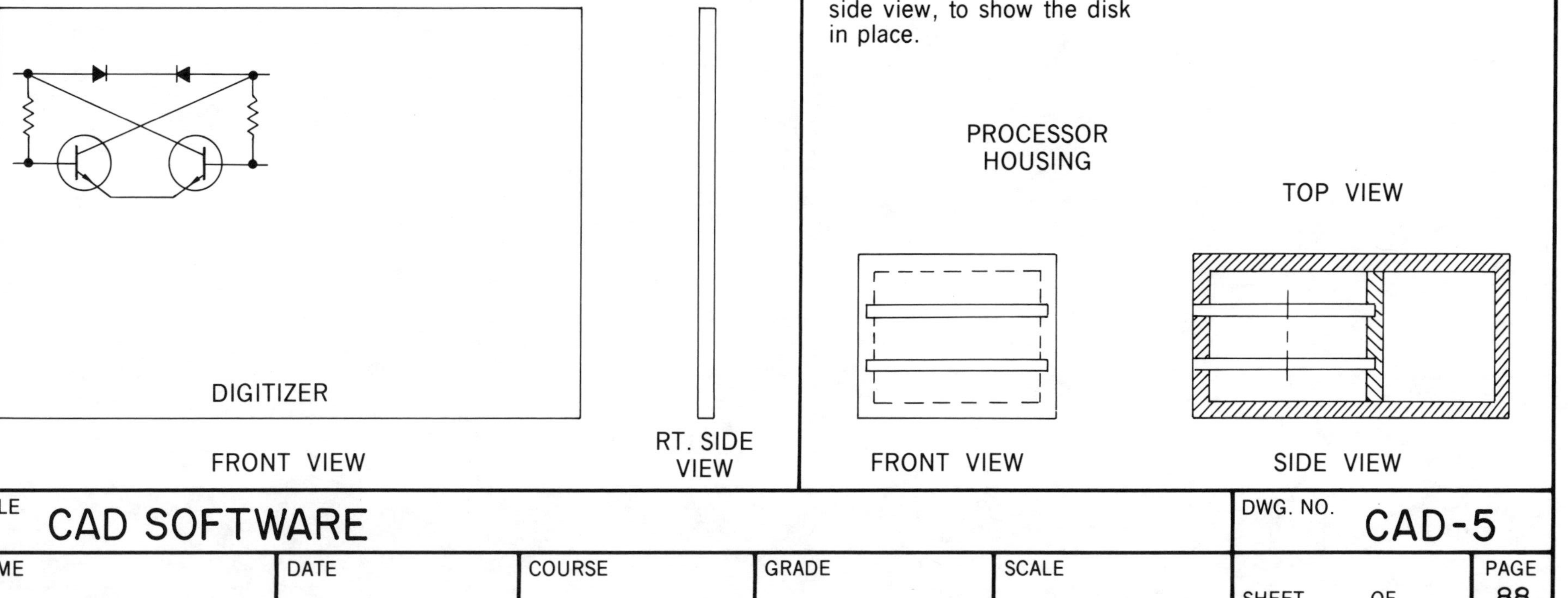

TITLE CAD SOFTWARE					DWG. NO. CAD-5	
NAME	DATE	COURSE	GRADE	SCALE	SHEET OF	PAGE 88

CAD SOFTWARE AND CREATIVE MENUS

In most mainframe and mini CAD systems, the menu-on-digitizer board unit is available, but filing and recall can also be achieved through the two keyboards. The menus in Appendix I and J are each limited to 25 entries, A through Y, and are filed or recalled through the "alphanumeric" or "function" keyboards. It should be emphasized that any symbol or component can be drawn as needed, particularly if it is not yet included in the menus; however, it is far more efficient to recall an item from the menu "storage", than to draw as needed.

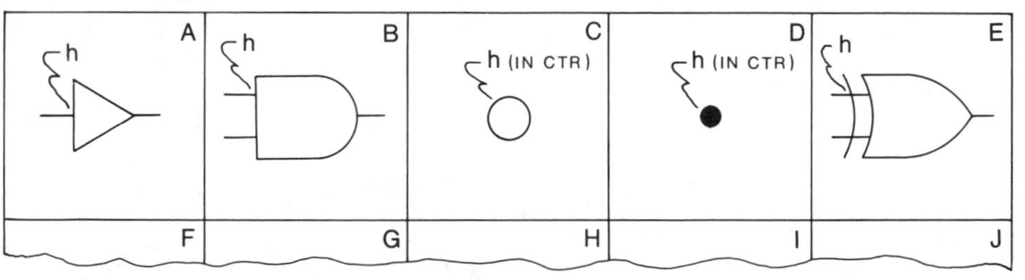

Fig. 1 Partial Appendix I

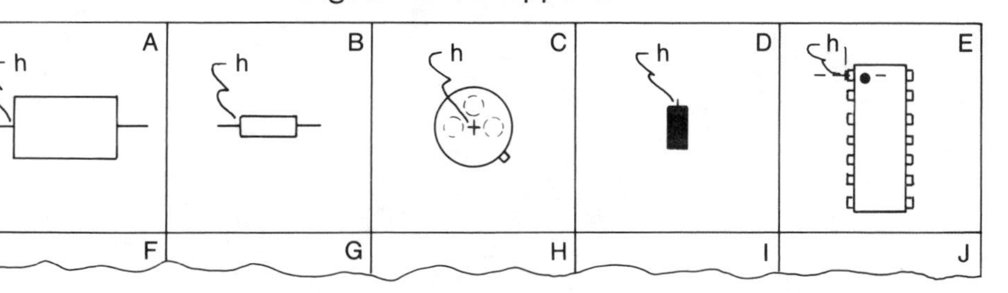

Fig. 2 Partial Appendix J (full scale)

Figures 1 and 2 are partial views of the menus I and J slightly modified, to indicate the "joining point" or "handle point" (h) where the symbol or component replaces the cursor (+) on the screen. Generally, the handle point (h) is at the left conductor entry/joining point; or upper left joint (for multiple entries); or circle center for circular symbols/components. The "h" is also the handle point where the cursor can "grip" the item to move it across the screen to another location, or about which the item can be rotated.

Rotation is also considered about the X, Y, and Z axes as well as the h point. For example, in CAD-3, the mirroring technique was also a rotation about the Y axis (℄). In Fig. 3, the rotations are either C or CC about the Z axis or about the X axis.

Y
90° C.C. (COUNTER CLOCKWISE) ABOUT Z AXIS
h
X
90°C. (CLOCKWISE) ABOUT Z AXIS
Z

Y
ORIGINAL POSITION
h
X

Y
180° ROTATION ABOUT X AXIS
h
X

Fig. 3 Rotations

Exercise: To draw the flip flop of page 37/84 or similar schematics by either mirroring *or* repetitive placement; complete (draw in) the partial I and J menus below by including the following rotational additions:

Symbols of Appendix I	For components of Appendix J
SQ. U = Capacitor, 90°C.C.	SQ. P = Diode, 180°C.C.
SQ. V = Resistor, 90°C.	SQ. Q = Transistor, 180° about X axis-hidden lines.
SQ. W = Transistor, 180° about the Y axis	SQ. R = Resistor, 90°C.
SQ. X = Diode, 180°C.C.	SQ. S = Diode, 90°C.
SQ. Y = Capacitor, 180°C.	SQ. T = Transistor, 180° about Z axis.

Indicate the new preferred "h" in all items below. If your instructor approves, you may include (draw in) the items below in Appendixes I and J.

Note: All original positions are shown in Appendixes I and J in solid (visible) lines.

SQUARES U-Y FOR SYMBOLS   SQUARES P-T FOR COMPONENTS

U	V	W	X	Y
P	Q	R	S	T

TITLE CAD SOFTWARE AND CREATIVE MENUS					DWG. NO. CAD–6	
NAME	DATE	COURSE	GRADE	SCALE	SHEET OF	PAGE 89

CAD KEYBOARD AND FUNCTION BOARD FAMILIARIZATION

In Appendix K there is a diagram of a keyboard and function board "as purchased" (Fig. 1) and as retitled (Fig. 2). This keyboard is the one which will be used in the lessons which follow. The retitled keyboards are used to facilitate learning, as shown in Appendix K abbreviations: STR1 = start 1 = press (1) = (1) and is used to start a line, a delete, a crosshatch, etc. SLD2 = solid 2 = press (2) = (2) and is used to end a solid line, do a solid rectangle, etc. See appendix for remaining interpretations. In this system the function board and keyboard number keys are interchangeable.

You have been asked to redesign a small keyboard/function board below. Your design should include keys .50″ high × .50″ wide, full scale, with space bar .50″ high × 5.50″ long. The keys are to be arranged with 5 rows × 15 keys/row, or 75 keys — not counting the space bar row. Use the revised key titles of Fig. 2 and relabel all keys approximately in their present locations starting at left and all others as "spare" (label: SPR) on the right side or top center — to be used later. Do all lettering 1/8″ high — neatly. For convenience, the abbreviations at right may be used.

Note: If this exercise is done with CAD, the 75 keys can be laid out automatically (step and repeat process) and the lettering would be done by typewriter (text edit).

ABBREVIATIONS USED

BACK SPACE = BKSP
BREAK = BRK
CLEAR = CLR
DISPLAY = DSPL
DELETE = DLĒT
ENTER = ENTR
PAUSE = PAUS
RETURN = RET
SCROLL = SCRL
SHIFT = SHFT
SPARE = SPR

PF1	PF2	PF3	▲
7	8	9	▼
FIG 4	5	6	▶
STR 1	SLD 2	BKN 3	◀
ϕ	.	ENTR	SPR

KEYBOARD

FUNCTION BOARD

TITLE CAD KEYBOARD AND FUNCTION BOARD FAMILIARIZATION					DWG. NO. CAD-7	
NAME	DATE	COURSE	GRADE	SCALE FULL	SHEET OF	PAGE 90

COMMON TASKS (ROUTINES) FOR CAD — PREPARATION OF ELECTRONIC DRAWINGS

Since every CAD system has its own procedures, it is difficult to generalize. However, one system* description may give one an idea of the routines involved for various capabilities. In all routines, including start-up, we will assume the proper software (disks) has been inserted and all power connections are properly hooked up. It should also prove helpful to inspect the two-keyboards diagram of Appendix K.

The next several pages of lessons will be devoted to familiarization with CAD capabilities through use of the most common tasks and routines needed to generate drawings with the particular CAD system* used in this workbook. See Appendixes L through T for tasks/routines and explanations.

An example of how Appendix L could be used is shown below: Assume you were going to start the CAD system on the 19th of December 1985, at 9:30AM, with "A" size paper, activating screen, grid, and cursor. Finally, illustrate the CRT image that would appear on the screen if you drew a line from point P3 (12,7) to point P4 (19,10).

[1] Turn on power

[2] Type 19 Dec 1985 and press RET

[3] Type 9:30AM and press RET

[4] Press RET twice

[5] Type @ASIZE and press RET

[6] Press TAB (Screen)

[7] Type GR and press RET

[8] press cursor

To draw the line after task 8 was completed, proceed with task number 9 as follows:

[9] Press 1 at 1st pt. P3 (at X = 12, Y = 7) JS to 2nd pt. P4 (at X = 19, Y = 10) and press 2.

This creates a SOLID LINE between the two points as shown in Fig. 1.

Note: The rectangle of Fig. 1, including P1 and P2, was generated by using Task #12, Appendix M-Rectangle Solid.

GRID
(TASK NO. 7)
P2 (21,13½) = upper rt. corner of "A" size sheet
P4
P3
P1 (10,5) = lower left corner of "A" size sheet (8½ × 11)
Y
X
0
10
20
30

Fig. 1

*Andromeda Systems Inc. (See preface.)

Exercise 1. In column below, write in a start-up task (routine) using current date and time. Indicate "B" size paper; activate screen, grid, and cursor.

TASK NO.	TASKS & ROUTINES	TASK NO.	TASKS & ROUTINES
[1]	TURN ON POWER	[6]	PRESS
[2]	TYPE	[7]	
[3]		[8]	
		[9]	Complete exercise 2 and routine [9] below.
[4]			
[5]			

Exercise 2. In the space below (Fig. 2), complete the started drawing of the image you programmed above (exercise 1, at 1/10 scale), and indicate how you would create a line from pt. P3 (at X = 15, Y = 10) to P4 (at X = 22, Y = 13). Also draw the "B" size sheet at P1 and P2.

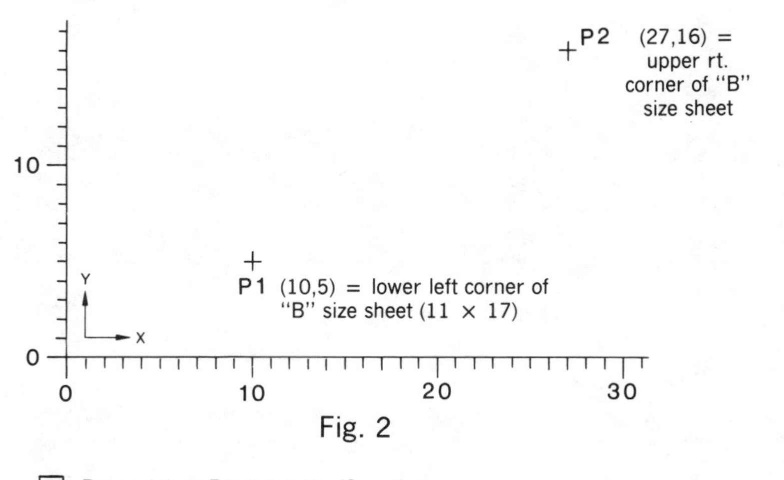

Fig. 2

[9] Press (1) at P3 (15,10), JS and press ______

TITLE COMMON TASKS AND ROUTINES FOR CAD

DWG. NO. CAD-8

NAME	DATE	COURSE	GRADE	SCALE	SHEET OF	PAGE 91

CAD TASKS AND PROGRAM WORK SHEETS

In the next several lessons, you will have to use Appendixes L through T plus a "program work sheet" to do the exercises. The work sheet is a simple step-by-step sequence of commands or tasks used to execute drawings with the CAD system. As an example, in CAD 8 (page 91) the solid line P3-P4 of Fig. 1 was executed by Task No. [9] ; but the **notation** could more simply be shown by the **work sheet** at right.

PROGRAM WORK SHEET		
Solid line	JS to 12, 7	PRS 1
	JS to 19, 10	PRS 2

Exercise 1.
Rewrite Task No. [10] , started at right, exactly as shown in Appendix L, including the explanations and sketch (1/8 lettering).

TASK NO.	TASKS & ROUTINES	EXPLANATION
[10]	LINE — CONTINUES (IGM) 10. PRESS 1	THIS
[11]	LINE — BROKEN (HIDDEN)	

Exercise 2. Follow the example used in Appendix M of TASK NO. 17 (Recall Electronic Symbols) to complete the started routine below. "Call-up" a diode, instead of an amplifier, whose handle (h) is at P1 (5,6). Draw the symbol in Fig. 1.

In the space below, complete the routine to recall a **Diode**.

PRESS R/R

JS TO 5, 6

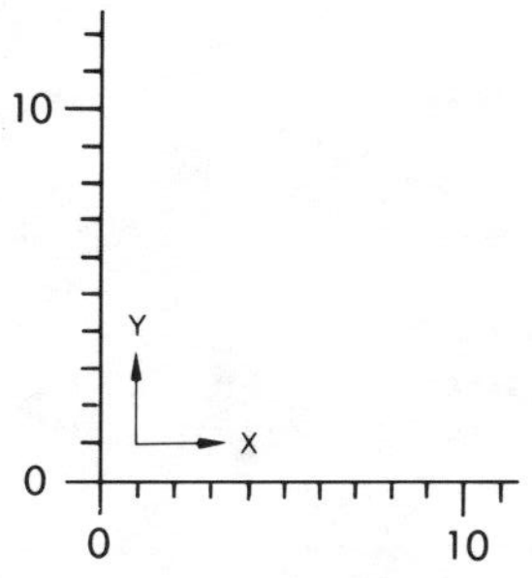

Fig. 1

Exercise 3. (See Appendix L.)
Complete note IV: TO GO

Exercise 4. The two ICs, U1 and U2, in Fig. 2 are interconnected as follows:

Front (Component side) Solid lines
- (a) U1-9 to U2-6
- (b) U1-11 to U2-5
- (c) U1-12 to U2-3

Back Dash lines (broken lines)
- (d) U1-10 to U2-2

Complete the started program work sheet below, showing the interconnections of the 4 conductors a, b, c, and d.

PROGRAM WORK SHEET		
SOLID LINE a	JS TO	
U1-9 TO U2-6	JS TO	
	JS TO 9.5, 5	PRS 1
SOLID LINE b	JS TO 12, 5	PRS 2
U1-11 TO U2-5	JS TO 12, 4	PRS 2
	JS TO 13.5, 4	PRS 2
SOLID		
BROKEN		

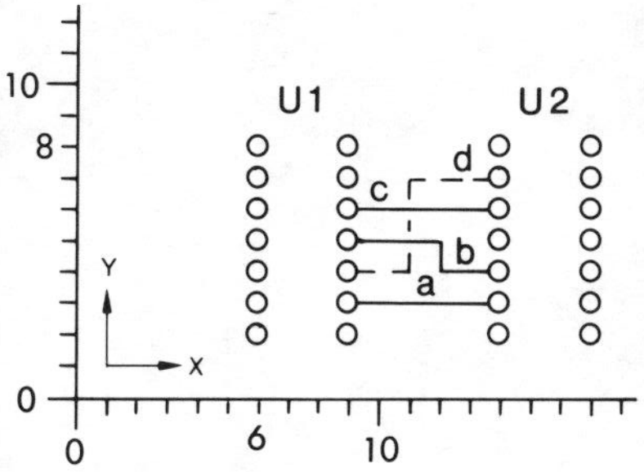

Fig. 2

TITLE CAD TASKS AND PROGRAM WORK SHEETS					DWG. NO. CAD-9	
NAME	DATE	COURSE	GRADE	SCALE	SHEET OF	PAGE 92

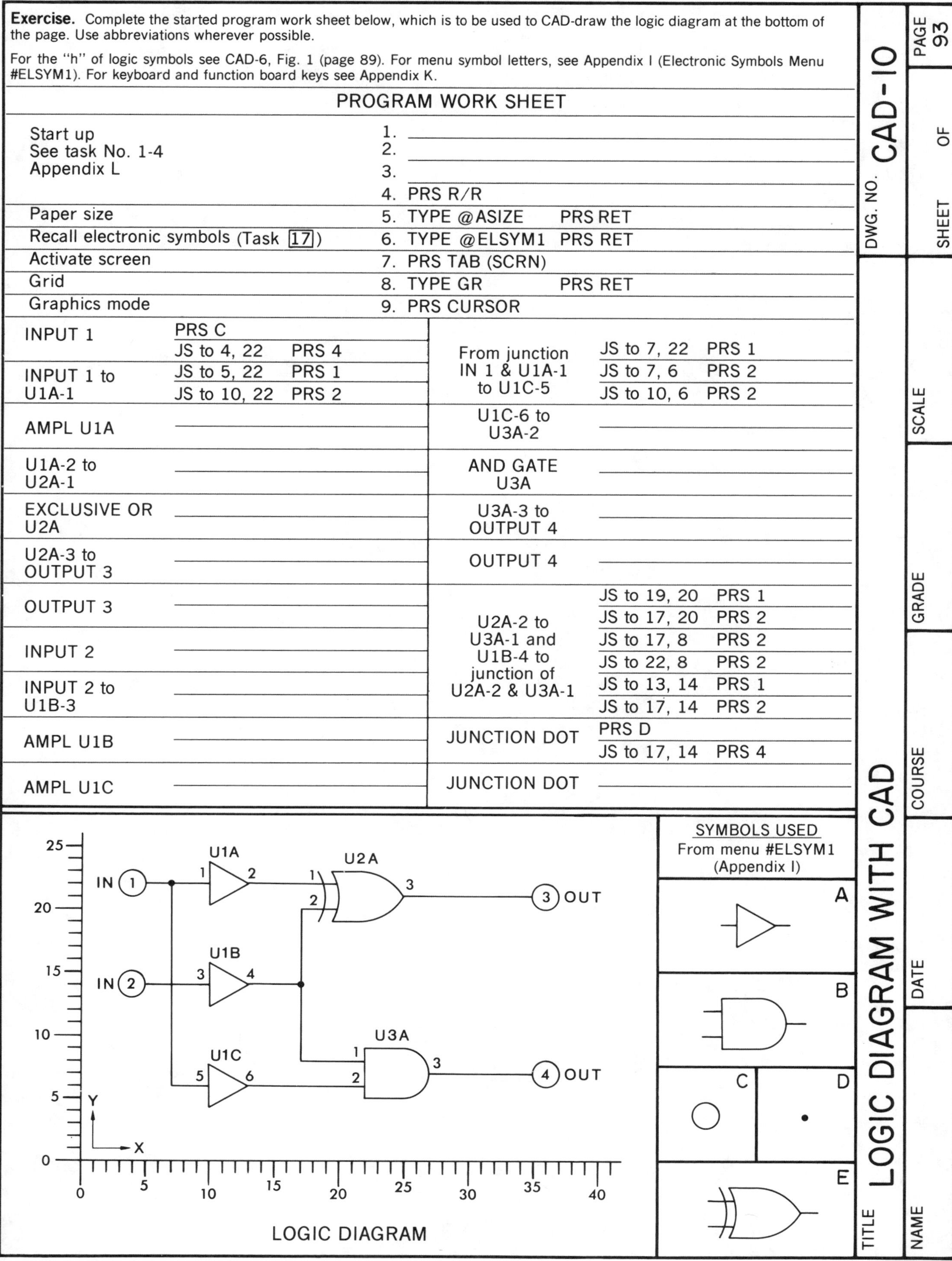

Exercise. Complete the started program work sheet below, which is to be used to CAD-draw the logic diagram at the bottom of the page. Use abbreviations wherever possible.

For the "h" of logic symbols see CAD-6, Fig. 1 (page 89). For menu symbol letters, see Appendix I (Electronic Symbols Menu #ELSYM1). For keyboard and function board keys see Appendix K.

PROGRAM WORK SHEET

Start up See task No. 1-4 Appendix L	1. ______ 2. ______ 3. ______ 4. PRS R/R
Paper size	5. TYPE @ASIZE PRS RET
Recall electronic symbols (Task [17])	6. TYPE @ELSYM1 PRS RET
Activate screen	7. PRS TAB (SCRN)
Grid	8. TYPE GR PRS RET
Graphics mode	9. PRS CURSOR

INPUT 1	PRS C JS to 4, 22	 PRS 4	From junction IN 1 & U1A-1 to U1C-5	JS to 7, 22 JS to 7, 6 JS to 10, 6	PRS 1 PRS 2 PRS 2
INPUT 1 to U1A-1	JS to 5, 22 JS to 10, 22	PRS 1 PRS 2			
AMPL U1A	______		U1C-6 to U3A-2	______	
U1A-2 to U2A-1	______		AND GATE U3A	______	
EXCLUSIVE OR U2A	______		U3A-3 to OUTPUT 4	______	
U2A-3 to OUTPUT 3	______		OUTPUT 4	______	
OUTPUT 3	______		U2A-2 to U3A-1 and U1B-4 to junction of U2A-2 & U3A-1	JS to 19, 20 JS to 17, 20 JS to 17, 8 JS to 22, 8 JS to 13, 14 JS to 17, 14	PRS 1 PRS 2 PRS 2 PRS 2 PRS 1 PRS 2
INPUT 2	______				
INPUT 2 to U1B-3	______				
AMPL U1B	______		JUNCTION DOT	PRS D JS to 17, 14	 PRS 4
AMPL U1C	______		JUNCTION DOT	______	

LOGIC DIAGRAM

SYMBOLS USED

From menu #ELSYM1 (Appendix I)

A

B

C

D

E

DWG. NO. CAD-10

PAGE 93

SHEET OF

SCALE

GRADE

COURSE

DATE

NAME

TITLE LOGIC DIAGRAM WITH CAD

Exercise

Complete the started program work sheet at right, which is to be used to CAD-draw the PCB diagram below—as in CAD 10 (page 93). The first nine steps of CAD-10 will be assumed similarly completed here with step 6 rewritten as: 6. Type @ELCOM1 PRS RET. Other new tasks used in this exercise are: [11] Broken lines and [26] Step and repeat.

SYMBOLS USED
From menu #ELCOM1
(Appendix J)

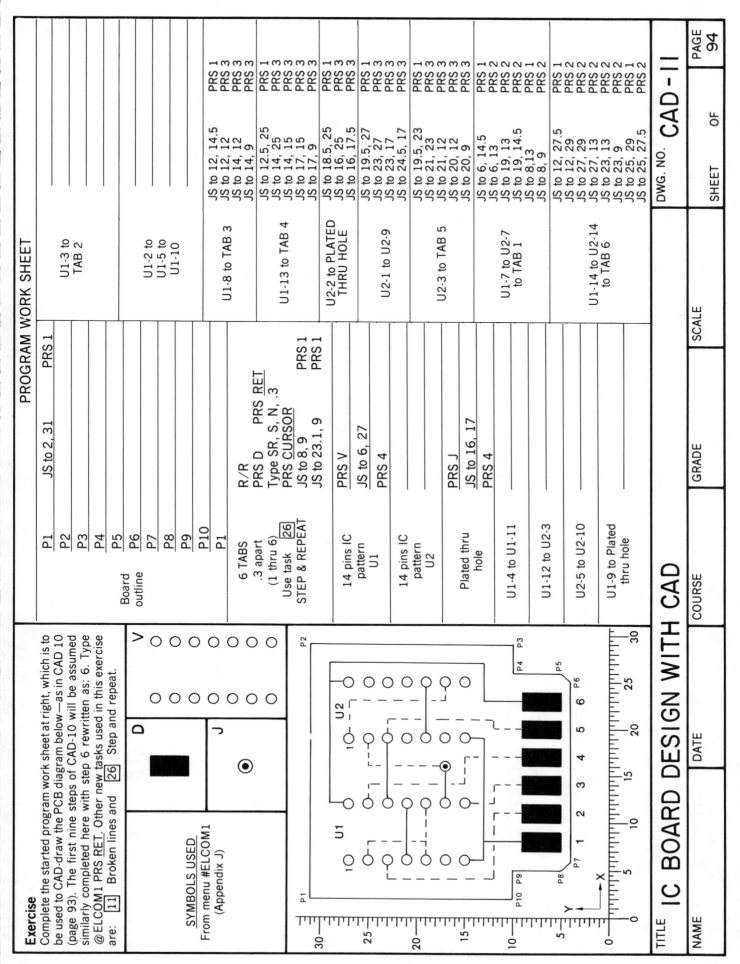

D J  V

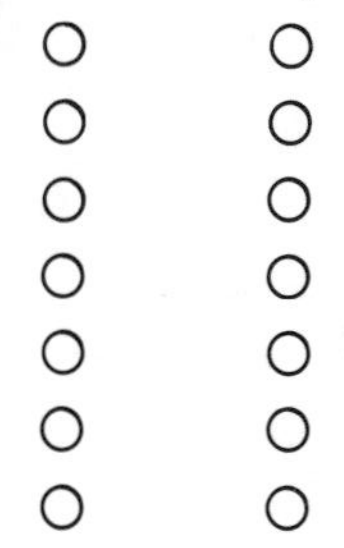

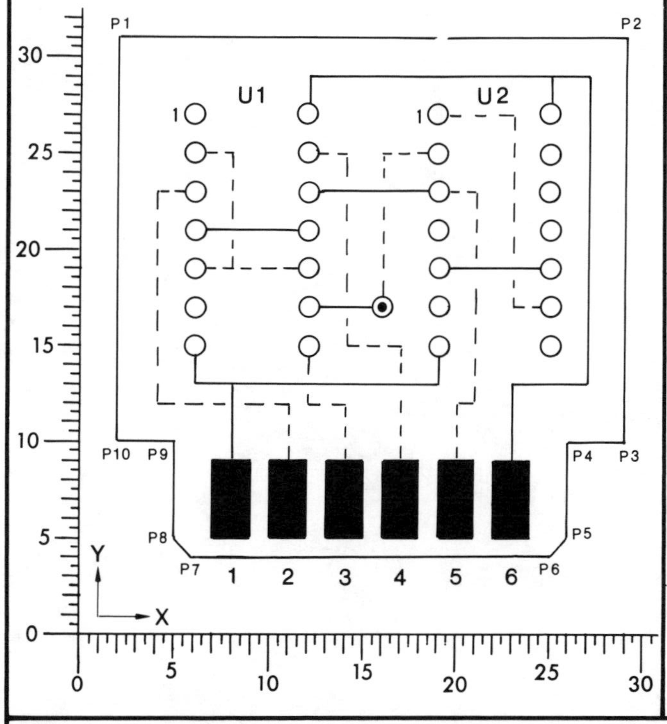

U1 U2 1 1 P1 P2 P3 P4 P5 P6 P7 P8 P9 P10 1 2 3 4 5 6 Y X

0 5 10 15 20 25 30

PROGRAM WORK SHEET

Task	Steps	
Board outline	P1 JS to 2, 31	PRS 1
	P2	
	P3	
	P4	
	P5	
	P6	
	P7	
	P8	
	P9	
	P10	
	P1	
6 TABS .3 apart (1 thru 6) Use task [26] STEP & REPEAT	R/R	
	PRS D PRS RET	
	Type SR, S, N, .3	
	PRS CURSOR	
	JS to 8, 9	PRS 1
	JS to 23.1, 9	PRS 1
14 pins IC pattern U1	PRS V	
	JS to 6, 27	
	PRS 4	
14 pins IC pattern U2		
Plated thru hole	PRS J	
	JS to 16, 17	
	PRS 4	
U1-4 to U1-11		
U1-12 to U2-3		
U2-5 to U2-10		
U1-9 to Plated thru hole		

Task	Steps	
U1-3 to TAB 2		
U1-2 to U1-5 to U1-10		
U1-8 to TAB 3	JS to 12, 14.5	PRS 1
	JS to 12, 12	PRS 3
	JS to 14, 12	PRS 3
	JS to 14, 9	PRS 3
U1-13 to TAB 4	JS to 12.5, 25	PRS 1
	JS to 14, 25	PRS 3
	JS to 14, 15	PRS 3
	JS to 17, 15	PRS 3
	JS to 17, 9	PRS 3
U2-2 to PLATED THRU HOLE	JS to 18.5, 25	PRS 1
	JS to 16, 25	PRS 3
	JS to 16, 17.5	PRS 3
U2-1 to U2-9	JS to 19.5, 27	PRS 1
	JS to 23, 27	PRS 3
	JS to 23, 17	PRS 3
	JS to 24.5, 17	PRS 3
U2-3 to TAB 5	JS to 19.5, 23	PRS 1
	JS to 21, 23	PRS 3
	JS to 21, 12	PRS 3
	JS to 20, 12	PRS 3
	JS to 20, 9	PRS 3
U1-7 to U2-7 to TAB 1	JS to 6, 14.5	PRS 1
	JS to 6, 13	PRS 2
	JS to 19, 13	PRS 2
	JS to 19, 14.5	PRS 2
	JS to 8,13	PRS 1
	JS to 8, 9	PRS 2
U1-14 to U2-14 to TAB 6	JS to 12, 27.5	PRS 1
	JS to 12, 29	PRS 2
	JS to 27, 29	PRS 2
	JS to 27, 13	PRS 2
	JS to 23, 13	PRS 2
	JS to 23, 9	PRS 2
	JS to 25, 29	PRS 1
	JS to 25, 27.5	PRS 2

TITLE IC BOARD DESIGN WITH CAD

DWG. NO. CAD-11

NAME | DATE | COURSE | GRADE | SCALE | SHEET OF

Exercise. Follow the CAD program work sheet below and complete the started hybrid circuit layout drawing (Fig. 1). After the layout is completed, identify the components (R1, Q1, CR1, etc.) manually, using 1/8″ lettering rather than CAD text. (References used include Tasks #9, 10, 11, 12, and ELSYM1 menu.)

CAD PROGRAM WORK SHEET

1. Turn on power
2. Type in date PRS RET
3. Type in time PRS RET
4. PRS R/R
5. Type @ASIZE PRS RET
6. PRS TAB (SCRN)
7. Type GR PRS RET

For the next eight "rectangle-solid" entries below, use command: PRS R PRS CURSOR. Now you can draw components as per coordinates indicated below.

Resistor R1	JS to 5,53	PRS	2
	JS to 11,41	PRS	2
Diode CR1 (Cathode)	JS to 5,31	PRS	2
	JS to 11,25	PRS	2
Diode CR1 (Anode)	JS to 7,29	PRS	2
	JS to 9,27	PRS	2
Capacitor C1	JS to 5,20	PRS	2
	JS to 11,5	PRS	2
Transistor Q1 (Collector)	JS to 15,47	PRS	2
	JS to 25,37	PRS	2
Transistor Q1 (Emitter)	JS to 19,43	PRS	2
	JS to 21,41	PRS	2
Transistor Q1 (Base)	JS to 16,40	PRS	2
	JS to 18,38	PRS	2
Resistor R2	JS to 15,29	PRS	2
	JS to 25,11	PRS	2

You are now ready to draw conductors (next six entries).

Conductor #1 Solid and broken lines (top)	JS to 28,54	PRS	1
	JS to 4,54	PRS	2
	JS to 4,50	PRS	2
	JS to 5,50	PRS	2
	JS to 5,50	PRS	1
	JS to 11,50	PRS	3
	JS to 11,50	PRS	1
	JS to 28,50	PRS	2
Conductor #2 Solid and broken lines (2nd from top left)	JS to 11,44	PRS	1
	JS to 12,44	PRS	2
	JS to 12,35	PRS	2
	JS to 4,35	PRS	2
	JS to 4,44	PRS	2
	JS to 5,44	PRS	2
	JS to 5,44	PRS	1
	JS to 11,44	PRS	3
Conductor #3 Solid and broken lines (3rd from top left)	JS to 11,17	PRS	1
	JS to 12,17	PRS	2
	JS to 12,32	PRS	2
	JS to 4,32	PRS	2
	JS to 4,17	PRS	2
	JS to 5,17	PRS	2
	JS to 5,17	PRS	1
	JS to 11,17	PRS	3
Conductor #4 Solid and broken lines (2nd from top right)	JS to 25,26	PRS	1
	JS to 26,26	PRS	2
	JS to 26,48	PRS	2
	JS to 14,48	PRS	2
	JS to 14,26	PRS	2
	JS to 15,26	PRS	2
	JS to 15,26	PRS	1
	JS to 25,26	PRS	3
Conductor #5 Solid and broken lines (3rd from top right)	JS to 28,14	PRS	1
	JS to 25,14	PRS	2
	JS to 25,14	PRS	1
	JS to 15,14	PRS	3
	JS to 15,14	PRS	1
	JS to 14,14	PRS	2
	JS to 14,10	PRS	2
	JS to 28,10	PRS	2
Conductor #6 Solid and broken lines (bottom)	JS to 28,8	PRS	1
	JS to 11,8	PRS	2
	JS to 11,8	PRS	1
	JS to 5,8	PRS	3
	JS to 5,8	PRS	1
	JS to 4,8	PRS	2
	JS to 4,4	PRS	2
	JS to 12,4	PRS	2
	JS to 12,6	PRS	2
	JS to 28,6	PRS	2
Jumper Q1-E	JS to 20,52	PRS	1
	JS to 20,42	PRS	2
Jumper Q1-B	JS to 17,39	PRS	1
	JS to 10,37	PRS	2
Jumper CR1 Anode	JS to 6,37	PRS	1
	JS to 8,28	PRS	2

PRS R/R
Type @ELSYM1 PRS RET PRS CURSOR
PRS D
(this is the DOT)

JS to 20,52	PRS	4
JS to 10,37	PRS	4
JS to 6,37	PRS	4

R1

Y X

0 5 10 15 20 25 30 35 40 45 50

Fig. 1
Portion of a
Hybrid Microcircuit Layout

TITLE HYBRID MICROCIRCUIT LAYOUT WITH CAD					DWG. NO. CAD-12	
NAME	DATE	COURSE	GRADE	SCALE	SHEET OF	PAGE 95

A new CAD operator made the mistake of CAD-drawing a PCB **full size** while using the ELCOM1 menu **2 × size terminals** (see Fig. 1). He used the left column below for his first program work sheet.

14 CLEAR SCREEN
1. PRS R/R
2. PRS C PRS RET
3. PRS Y PRS RET
4. Type DP PRS RET

7 ACTIVATE GRID
5. Type GR PRS RET

12 RECTANGLE-SOLID
6. PRS R PRS CURSOR
7. JS to P1 (2.5, 2.5) PRS (2)
8. JS to P2 (4.5, 4.7) PRS (2)

(small PCB w/o terminals appears)

26 STEP AND REPEAT
9. PRS R/R
10. Type @ELCOM1 PRS RET
11. PRS O PRS RET
12. Type SR, S, N, .2
13. PRS CURSOR
14. JS to P3 (2.8, 2.65) PRS (1)
15. JS to P4 (4.3, 2.65) PRS (1)

(Terminals now appear as in Fig. 1.)

Exercise 1. Complete the started work sheet below to correctly CAD-draw the PCB and terminals 2 × size as per Fig. 2.

14 CLEAR SCREEN
1. PRS R/R
2. PRS C PRS RET
3. ______
4. ______

7 ACTIVATE GRID
5. ______

12 RECTANGLE-SOLID
6. ______
7. JS to P5 (.7, .7) ______
8. JS to P6 (______) ______

(large PCB w/o terminals appears)

26 STEP AND REPEAT
9. ______
10. Type ______ ______
11. ______
12. ______
13. ______
14. JS to P7 (1.3, 1) PRS ( )
15. JS to P8 (______) ______

(Terminals now appear as in Fig. 2.)

Exercise 2. Fill in missing parenthetical values and terminals of Fig. 2, as per Exercise 1.

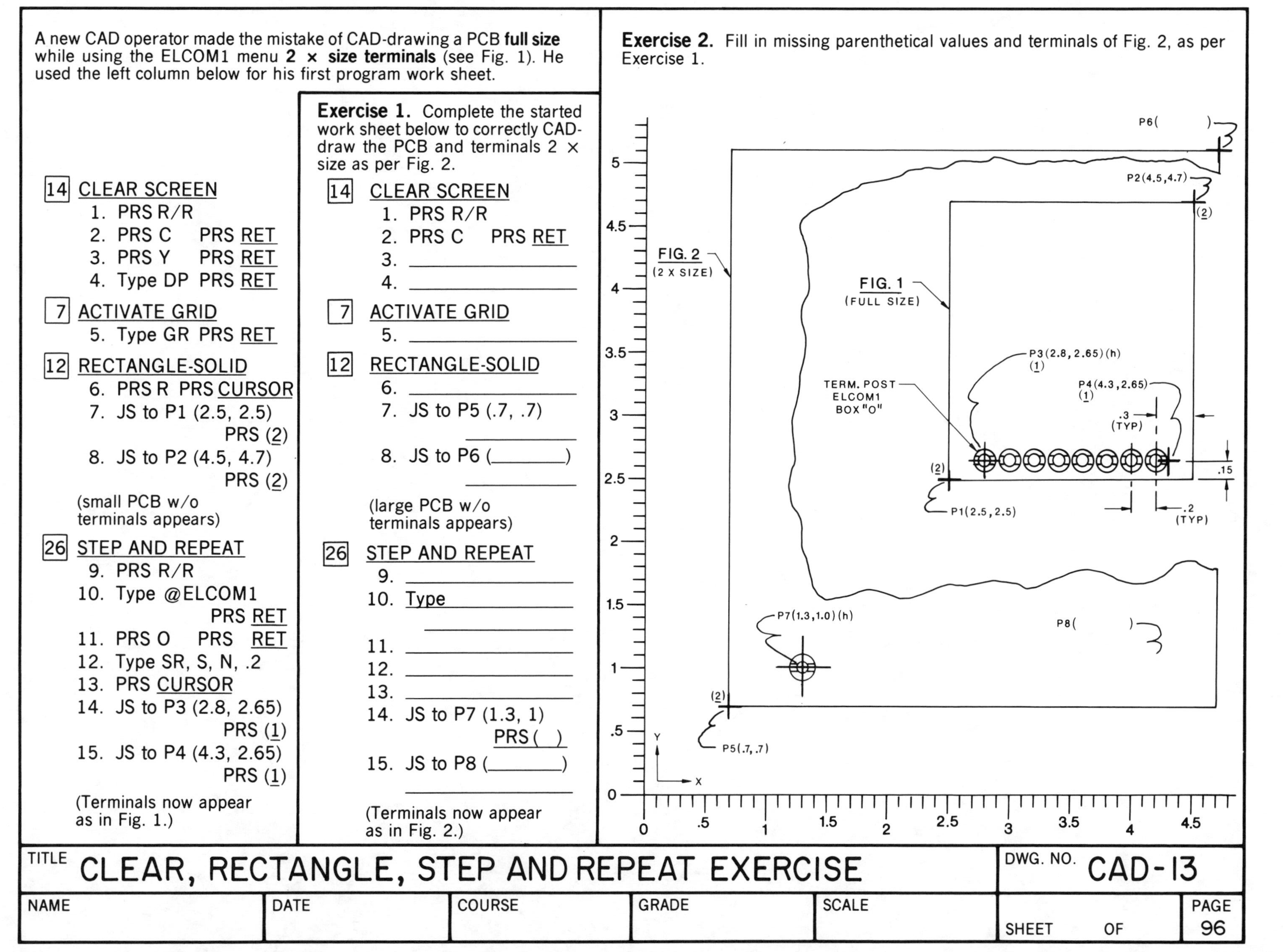

TITLE CLEAR, RECTANGLE, STEP AND REPEAT EXERCISE

DWG. NO. CAD-13

NAME	DATE	COURSE	GRADE	SCALE	SHEET OF	PAGE 96

Exercises. Let's assume you are employed with a supervisor who has been redrawing old "manually drafted" drawings with CAD for storage purposes. He has asked you to help him by finishing the right sides of the started sketches of the schematic and PCB, Figs. 1 and 2, by "mirroring." He has also decided to identify the left half of the schematic as MIRR03.SCH and the mirrored drawing as MIRR04.SCH. The left half and mirrored PCB are to be identified as MIRR03.PCB and MIRR04.PCB respectively. In the started columns below complete the routines required to mirror and save the two drawings. Use the "C" boxes of both the SCH and PCB menus for figure modification, since boxes A and B have already been occupied in Appendix Q (Task No. 27).

Exercise 1.
To mirror the schematic (Fig. 1) and place it in storage:
R/R
Type @PAGE 2 PRS RET
Press CURSOR
(CAD-redraw left half of Fig. 1), then:
R/R
Type
S, MIRR03.SCH PRS RET
Type
F, MIRR03.SCH, C PRS RET
Type
Type
Press CURSOR
JS to
PRS C PRS 4
mirrored Fig. 1 (both sides in full & solid) now appears on screen.

To store mirrored schematic:
R/R
Type
S, MIRR04.

Exercise 2.
To mirror the PCB of Fig. 2 and place it in storage:
R/R
Type ________
Press________
(CAD-redraw left half of Fig. 2), then:
____
Type
S, MIRR03.PCB
Type
Type
Type
Press CURSOR
JS to
mirrored PCB (both sides of Fig. 2) now appears on screen.

To store mirrored PCB:
R/R
Type

MIRR03.SCH

Fig. 1.
Schematic

LEFT SIDE | RT. SIDE

(0,0) ℄-FULCRUM

Exercise 3.
Complete right side sketch of mirrored Fig. 1.

Exercise 4.
Sketch the right side of the PCB at right in phantom lines to represent the mirror image as in the schematic above.

Fig. 2.
PCB

MIRR03.PCB

LEFT SIDE (0,0) ℄-FULCRUM RT. SIDE

TITLE SCHEMATIC–PCB MIRROR EXERCISE					DWG. NO. CAD-14	
NAME	DATE	COURSE	GRADE	SCALE	SHEET OF	PAGE 97

Complete the started "partial artwork" program worksheet of Fig. 1 so that the trim lines and register marks are crosshatch-filled-in. Afterwards, program erase the reduction note (as was a former PCB) and CAD write it along the P16–P17 line (Fig. 1*a*).

Finally, fill in manually everything you have programed—the reduction note may be crossed out and rewritten in its new position, above the underlining shown in Fig. 1*a*.

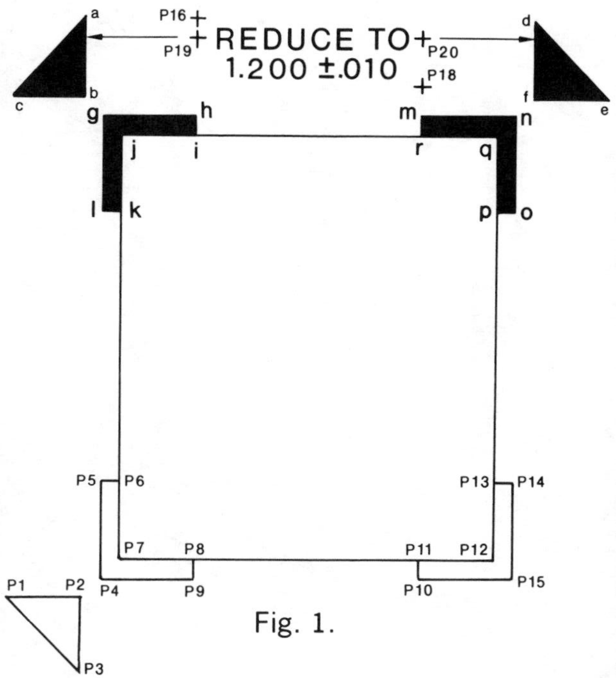

REDUCE TO
1.200 ±.010

a b c d e f g h i j k l m n o p q r P1 P2 P3 P4 P5 P6 P7 P8 P9 P10 P11 P12 P13 P14 P15 P16 P18 P19 P20

Fig. 1.

a b c d e f g h i j m n q r P16 P17

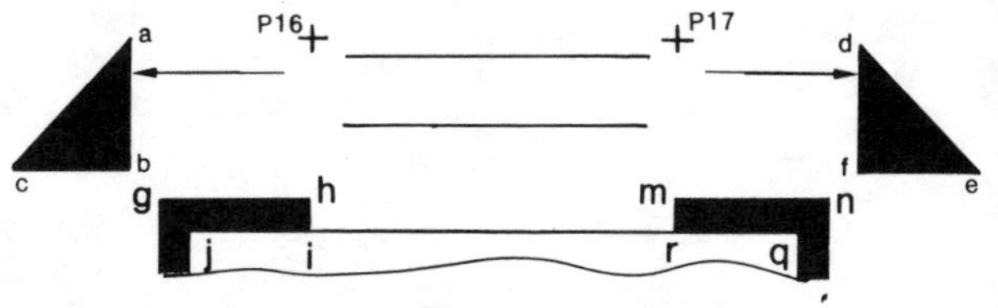

Fig. 1*a*.

Exercise 1.
Master artwork of Fig. 1
(Tasks 1-8 completed).

[24] CROSSHATCH FILL-IN
1. R/R
2. Type CH,NET3.PTN.01
3. Press CURSOR

(For top left register mark)
4. JS to Pa PRS (1)
5. JS to Pb PRS (2)
6. JS to Pc PRS (2)
7. Press R/R

(For top right register mark)
8. JS to Pd PRS (1)
9. JS to Pe PRS (2)
10. JS to Pf PRS (2)
11. Press R/R
12. (CAD-draw arrows with puck or joystick)

[30] TEXT (Lettering: reduction note)
13. R/R
14. Type T, CTF, STDFNT, .12, .5, 2, 0,1
15. Press CURSOR
16. JS to P19 PRS (1)
17. JS to P20 PRS (1)
18. Type REDUCE TO 1.200 ± .010
19. R/R

To erase a former PCB proceed as follows:

[20] ERASE (Pg – P15)
20. R/R
21. Type DEPRS CURSOR
22. JS to Pg PRS (3)
23. JS to P15 PRS (1)
24. R/R
25. Type PE PRS RET

(For the new Fig. 1)

[12] RECTANGLE-SOLID (P7–Pq)
26. R/R
27. Press R PRS CURSOR
28. JS to P7 PRS (2)
29. JS to Pq PRS (2)

[24] CROSSHATCH FILL-IN (Trim lines)
30. R/R
31. ______
32. ______

(Top left trim line)
33. JS to Pg
34. JS to Ph
35. JS to Pi
36. JS to Pj
37. ______
38. ______
39. R/R

(Top rt. trim line)
40. JS to Pm
41. JS to Pn
42. ______
43. ______
44. ______
45. JS to Pr PRS (2)
46. R/R

(Bottom left trim line)
47. JS to P5
48. JS to P6
49. JS to P7 PRS (2)
50. ______
51. ______
52. ______
53. R/R

(Bottom rt. Trim line)*
54. JS to P13
55. JS to P14
56. JS to P15
57. ______
58. ______
59. ______
60. R/R

(Bottom left register mark)
61. JS to P1
62. JS
63 ______
64. ______

(Erase old reduction note)

[20] ERASE (P16–P18)
65. R/R
66. Type
67. JS to
68. JS to
69. ______
70. Type

[30] TEXT (relocated reduction note—Fig. 1*a*)
71. ______
72. Type ______
73. ______
74. JS
75. ______
76. ______
77. R/R

*NOTE: Frequently used arrows, trim lines, register marks, etc., should be recallable via an additional menu (e.g., ELCOM2).

TITLE CAD-ARTWORK					DWG. NO. CAD-15	
NAME	DATE	COURSE	GRADE	SCALE	SHEET OF	PAGE 98

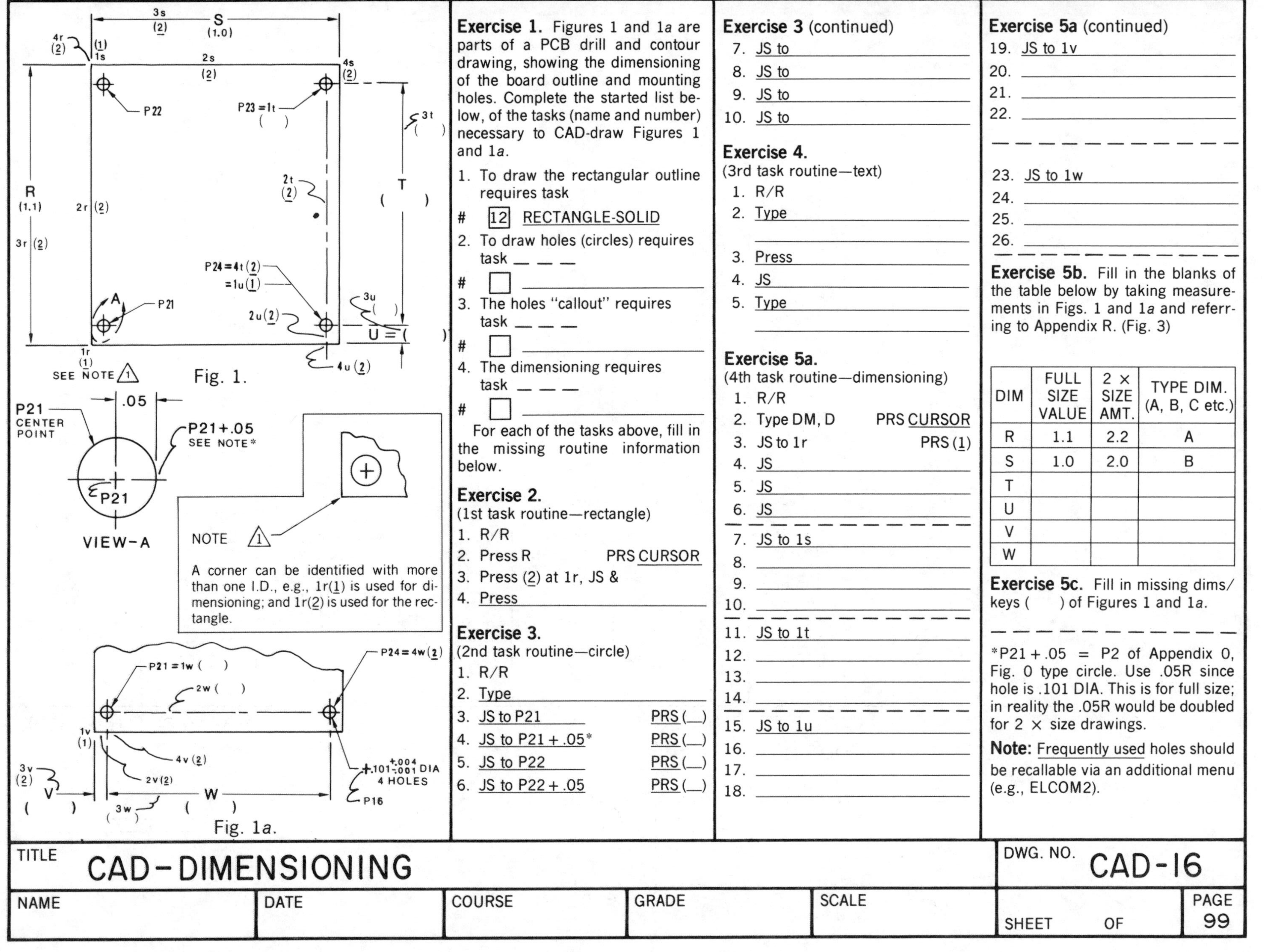

Fig. 1.

NOTE 1

A corner can be identified with more than one I.D., e.g., 1r(1) is used for dimensioning; and 1r(2) is used for the rectangle.

Fig. 1a.

Exercise 1. Figures 1 and 1*a* are parts of a PCB drill and contour drawing, showing the dimensioning of the board outline and mounting holes. Complete the started list below, of the tasks (name and number) necessary to CAD-draw Figures 1 and 1*a*.

1. To draw the rectangular outline requires task

   \# [12] RECTANGLE-SOLID

2. To draw holes (circles) requires task _ _ _

   \# [ ] ________

3. The holes "callout" requires task _ _ _

   \# [ ] ________

4. The dimensioning requires task _ _ _

   \# [ ] ________

For each of the tasks above, fill in the missing routine information below.

Exercise 2.
(1st task routine—rectangle)

1. R/R
2. Press R PRS CURSOR
3. Press (2) at 1r, JS &
4. Press ________

Exercise 3.
(2nd task routine—circle)

1. R/R
2. Type ________
3. JS to P21 PRS (_)
4. JS to P21 + .05* PRS (_)
5. JS to P22 PRS (_)
6. JS to P22 + .05 PRS (_)

Exercise 3 (continued)

7. JS to ________
8. JS to ________
9. JS to ________
10. JS to ________

Exercise 4.
(3rd task routine—text)

1. R/R
2. Type ________
3. Press ________
4. JS ________
5. Type ________

Exercise 5a.
(4th task routine—dimensioning)

1. R/R
2. Type DM, D PRS CURSOR
3. JS to 1r PRS (1)
4. JS ________
5. JS ________
6. JS ________

---

7. JS to 1s
8. ________
9. ________
10. ________

---

11. JS to 1t
12. ________
13. ________
14. ________

---

15. JS to 1u
16. ________
17. ________
18. ________

Exercise 5a (continued)

19. JS to 1v
20. ________
21. ________
22. ________

---

23. JS to 1w
24. ________
25. ________
26. ________

Exercise 5b. Fill in the blanks of the table below by taking measurements in Figs. 1 and 1*a* and referring to Appendix R. (Fig. 3)

DIM	FULL SIZE VALUE	2 × SIZE AMT.	TYPE DIM. (A, B, C etc.)
R	1.1	2.2	A
S	1.0	2.0	B
T			
U			
V			
W			

Exercise 5c. Fill in missing dims/keys ( ) of Figures 1 and 1*a*.

---

*P21 + .05 = P2 of Appendix O, Fig. O type circle. Use .05R since hole is .101 DIA. This is for full size; in reality the .05R would be doubled for 2 × size drawings.

Note: Frequently used holes should be recallable via an additional menu (e.g., ELCOM2).

TITLE CAD-DIMENSIONING					DWG. NO. CAD-16	
NAME	DATE	COURSE	GRADE	SCALE	SHEET OF	PAGE 99

CAD-RECTIFICATION OF AN OLD DRAWING

A new electronics CAD operator was given an old PCB drawing to redo because the Q1-E conductor was shorting-out at Q1-B. He was asked to CAD-redraw the sketched portion shown in Fig. 1, so as to eliminate the shorting-out problem.

The operator decided to proceed along the following series of goal-routines:

I. Recall (load) the PCB from storage (assumed I.D. #PCB006.DWG).

II. Delete Q1-E conductor (at P1 and P2).

III. CAD-redraw Q1-E conductor with sufficient Q1-B clearance — from Q1-E to P3,P4,P5, and P6 (see Fig. 2).

Exercise 1. Complete the started routine for the above goal I: Recall (load) the PCB (assume I.D. #PCB006.DWG).

I. [16] RECALL (LOAD) A DWG.
   1. Press R/R
   2. ____________
   3. ____________

After step I-3, a full size PCB appeared on the screen. The operator decided to make the drawing 2 × size because it was too small for modification.

Exercise 2. Write the routine necessary to increase PCB006.DWG to 2 × size.

A. [ ] ____________
   1. R/R
   2. ____________

Apparently 2 × size (Fig. 1) was also not large enough, so the operator decided to "zoom in" on Q1.

Exercise 3. Write the routine necessary to "zoom in" on Q1 using P7-P8 (Fig. 2).

B. [ ] ____________
   1. R/R
   2. ____________
   3. ____________
   4. ____________

At this point the operator could see that if he allowed sufficient clearance for Q1-B, there would not be sufficient clearance for Q1-C. He then decided to erase Q1-C and relocate it at P9.

Exercise 4. Write the routine necessary to erase Q1-C using P10-P11 before relocating it to P9 (Fig. 3).

C. [ ] ____________
   1. Press R/R
   2. ____________
   3. ____________
   4. ____________
   5. R/R
   6. ____________

After step C6 was executed, the operator noticed that there remained a couple of remnant conductors at P10-P13 (Fig. 3) and P11-P12 as well as the Q1-E conductor to delete.

The operator decided to delete all three conductors with one input command — at P10-P13 (Fig. 3), P11-P12, and (goal II) the Q1-E conductor at P10-P14 (instead of P1-P2).

(continue on page 101)

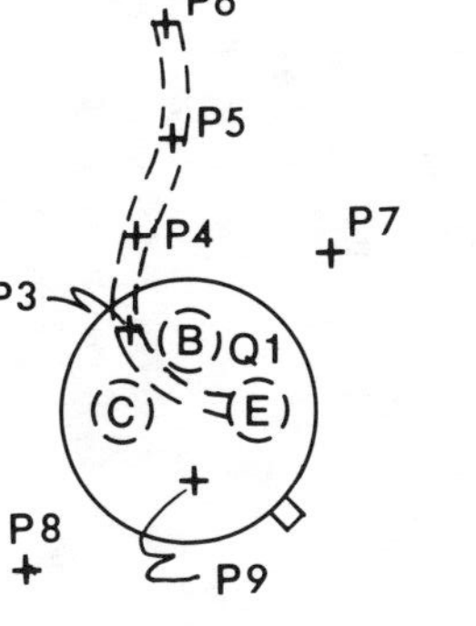

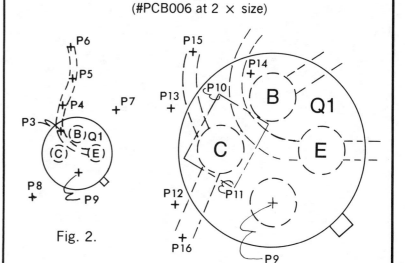

Fig. 1
(#PCB006 at 2 × size)

Fig. 2.

Fig. 3.

TITLE SHORT-OUT CAD RECTIFICATION | DWG. NO. CAD-17

NAME | DATE | COURSE | GRADE | SCALE | SHEET 1 OF 2 | PAGE 100

CAD-RECTIFICATION

(continued from page 100)

Exercise 5. Write the routine to delete the two remnant conductors and (goal II) Q1-E conductor with one input command, using the points and order suggested on the preceding page.

D. [ ] ______

1. Press R/R
2. ______
3. JS to P10 PRS (1)
4. ______
5. JS to P11
6. ______
7. ______
8. ______
9. ______
10. ______

After step D10 was executed, the Q1 appeared on the screen somewhat like the figure in Appendix T.

Exercise 6. Convert the started phantom view of Q1 below (Fig. 4) to the view that would appear after step D10.

Q1

Fig. 4

At this point the operator realized that in order to add a Q1-C pad at P9, Fig. 4 would have to be zoomed-out to its previously smaller 2 × scale size because the ELCOM1 menu pad is 2 × size.

Exercise 7. Write the routine to zoom-out Fig. 4 to its previously smaller 2 × scale size.

E. [ ] ______

1. ______
2. ______

Exercise 8. In the space below marked "Fig. 5" complete the view of Q1 that would appear after step E2.

Exercise 9. Write the routine to place new Q1-C pad on the P9 of Fig. 5.

F. [18] ______

1. Press R/R
2. Type ______
3. ______
4. ______
5. ______
6. ______

Exercise 10. Write the routine to replace: (a) double hidden line conductor from Q1-E to P3,P4,P5, and P6 of Fig. 2 (goal III) and similarly (b) conductor (P9)Q1-C to P13-P15 plus (c) conductor (P9) Q1-C to P16 of Fig. 3.

G. [ ] ______

(a) 1. Press R/R
2. ______
3. ______
4. ______
5. ______
6. ______
7. ______
8. ______

(b) 9. JS to (P9) Q1-C
10. ______
11. ______

(c) 12. ______
13. ______
14. Press R/R

The operator found that the new conductors were properly terminated—somewhat as in Fig. 1. Therefore he decided to save and plot the revised drawing (identified as #PCB007.DWG).

Exercise 11. Write the routine to save #PCB007. DWG.

H. [ ] ______

1. Press R/R
2. ______
3. ______

Exercise 12. Write the routine to plot #PCB007. DWG.

I. [32] ______

1. R/R
2. ______
3. ______

Exercise 13. Write the routine to clear the screen.

J. [ ] ______

1. R/R
2. ______
3. ______
4. ______

Exercise 14. In the space marked "Fig. 6" draw the view of Q1 that would appear on the "plot-out".

(indicate P9 crosshair)

Q1
(E)

Fig. 5

Q1
(E)

Fig. 6

TITLE SHORT-OUT CAD RECTIFICATION					DWG. NO. CAD-18	
NAME	DATE	COURSE	GRADE	SCALE	SHEET 2 OF 2	PAGE 101

ARTWORK MENUS

Your supervisor decided to make up an artwork menu that had all the company artwork aids which might be needed. He also wanted a commonly used PCB with its artwork aids included. (See menu at right.)

You have been given the task of completing this started menu.

Exercise 1.
Draw in the mirror image of register mark E in box H and label it properly. (**Hint:** see labeling of boxes A and D.)

Exercise 2.
Draw in the mirror image of lower left trim line F in box G and label it properly.

Exercise 3.
Fill-in all artwork aids requiring a "cross-hatch fill-in," including box I pads and tabs.

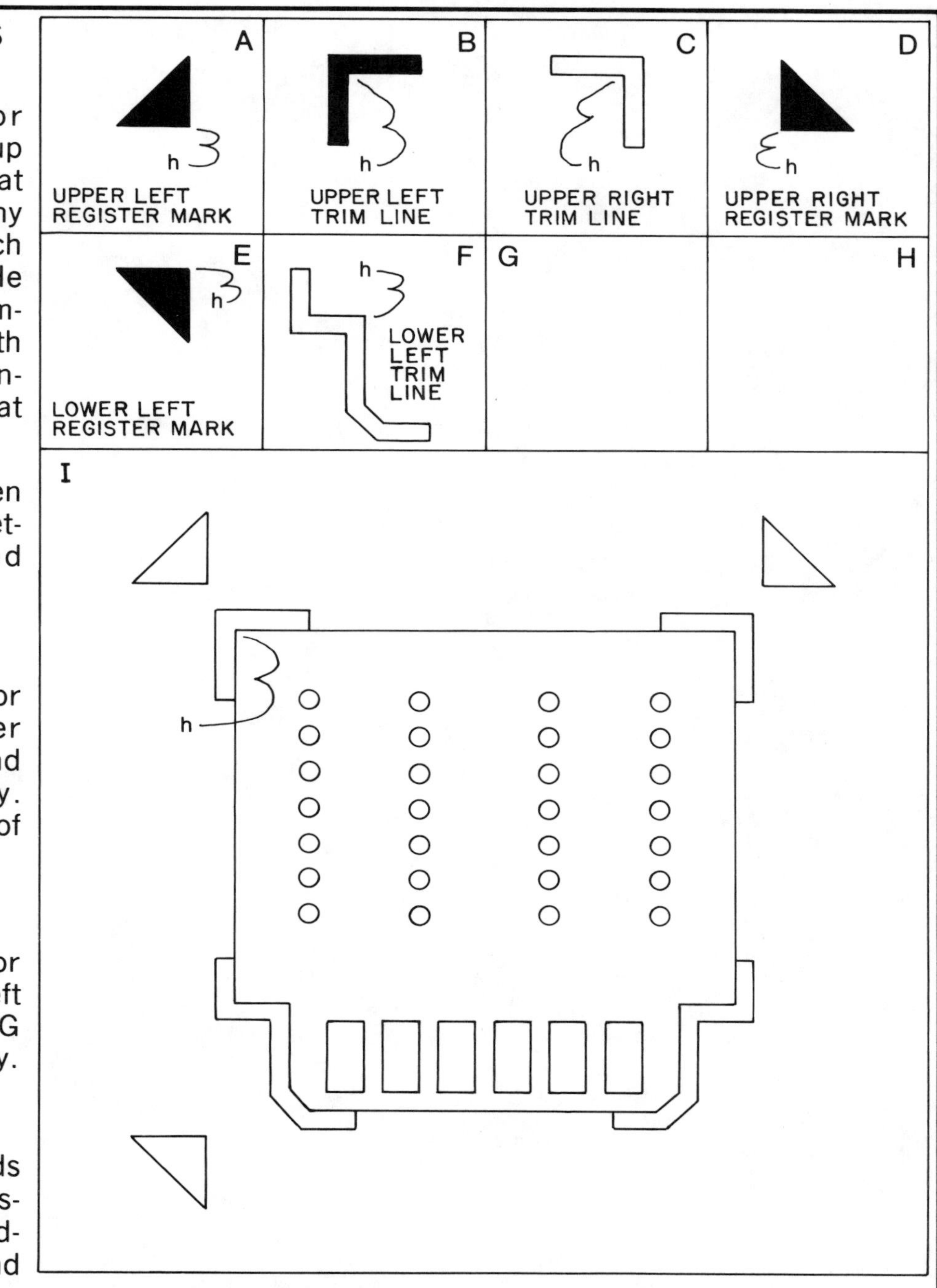

ARTWORK AIDS — MENU #ARTWK1
(Task no. 33 — similar to no. 18)

Exercise 4.
After box I has been crosshatch filled-in, label it "I.C. BOARD ARTWORK" in lower left corner of box I.

Exercise 5.
Insert a proper reduction note in box I with arrows.

DWG. NO. CAD-19 | PAGE 102 | SHEET OF

TITLE ARTWORK MENU

NAME | DATE | COURSE | GRADE | SCALE

I.C. CAD-Artwork: The Back Face Artwork of CAD-11 (page 94) shown at right, has already been CAD-drawn. You have been asked to complete the Front Face Artwork with CAD—partially done below.

Exercise 1. To recall the I.C. board artwork (I) of the artwork menu requires task—

☐ RECALL ARTWORK AIDS

1. Press R/R
2. Type @ARTWK1 PRS___
3. ________
4. JS to P1
5. Press ________
6. ________

Exercise 2. To superimpose a junction point (dot—box D) on plated through hole at pt. c and tool holes at pts. y and z requires task—

☐ ________

1. R/R________
2. Type ________
3. ________
4. JS to pt. c (16, 17)
5. Press D
6. ________
7. JS to pt. z (27, 11.5)
8. ________
9. ________
10. JS to pt. y (   )
11. ________
12. ________

REDUCE TO 1.500 ±.010

FRONT FACE

BACK FACE

When the BACK FACE conductors were CAD-drawn, the command "Type DB, .031, 2, 6" was used; "2", for solid center line and "6" for no. 6 pen which made .06" wide center lines.

Exercise 3. To CAD-draw double line solid conductors requires task—

☐ ________

1. Press R/R
2. Type ________
3. ________
4. JS to d PRS (__)
5. JS to e (__)
6. JS to f (__)
7. JS to g (__)
8. JS to h (__)
9. JS to i (__)
10. JS to j (__)
11. JS to c (__)
12. JS to k (__)
13. JS to l (__)
14. JS to m (__)
15. JS to n (__)
16. JS (__)
17. JS to p (__)
18. JS to q (__)
19. JS (__)
20. JS to s (__)
21. JS to t (__)
22. JS to (__)
23. JS (__)
24. JS to w (__)
25. JS (__)

Exercise 4. "Fill-in" the Front Face view above, somewhat like the Back Face view, to correspond with your program above: This includes fill-in of tabs, pads, and conductors (like fg: f━━g), etc.

TITLE I.C. BOARD, CAD-ARTWORK					DWG. NO. CAD-20	
NAME	DATE	COURSE	GRADE	SCALE 2 X SIZE	SHEET OF	PAGE 103

CAD-DRILL and Contour (CDC) drawing abbreviated preparations: (Abbreviated preparations are only a representative number of the required CAD-programed preparations—just to remind students of the tasks/routines needed without becoming repetitious.) The PCB at right, which was CAD-artworked on page 102 (CAD-19), had the reduction note, register marks, and trim lines erased and filed as "PCB011.DWG," before dimensions were drawn in. As a first step in a CDC dwg. preparation, indicate how this drawing would be recalled by completing the started program below.

Exercise 1. To recall Fig. 1 (W/O dims) requires task—

☐ ______

1. Press R/R
2. Type ______
3. ______

Exercise 2. To place dims "T" and "W", e.g., on dwg. requires task—

☐ ______

1. Press R/R
2. Type ______
3. JS to 1t ______ PRS (__)
4. JS to 2t ______ (__)
5. JS to ______ __
6. ______ __
7. JS to 1w ______ PRS (__)
8. ______ __
9. ______ __
10. ______ __

T is a(n) __ (A, B, C, etc.) type dim., whereas W is a(n) __ dim.

Exercise 3. To CAD-draw the rectangular hole chart using P1 and P2 requires task—

12 RECTANGLE-SOLID

1. Press R/R
2. ______
3. Press (__) at P1 JS
4. Press (__) at ______

After establishing the rectangular hole chart via P1 and P2, add horizontal and vertical lines inside the rectangle by using task no. ______. Omit programming this CAD-lines portion.

Exercise 4. To CAD-write the "Finish" note requires task—

☐ ______

1. Press R/R
2. Type ______
   ______
3. Press ______
4. JS ______
5. JS ______
6. Type FINISH: ______
   ______
   ______
   ______
7. R/R

Exercise 5. Crosshatch fill-in the PCB artwork manually, to illustrate what Fig. 1 would look like after having been CAD-artworked and modified as described above.

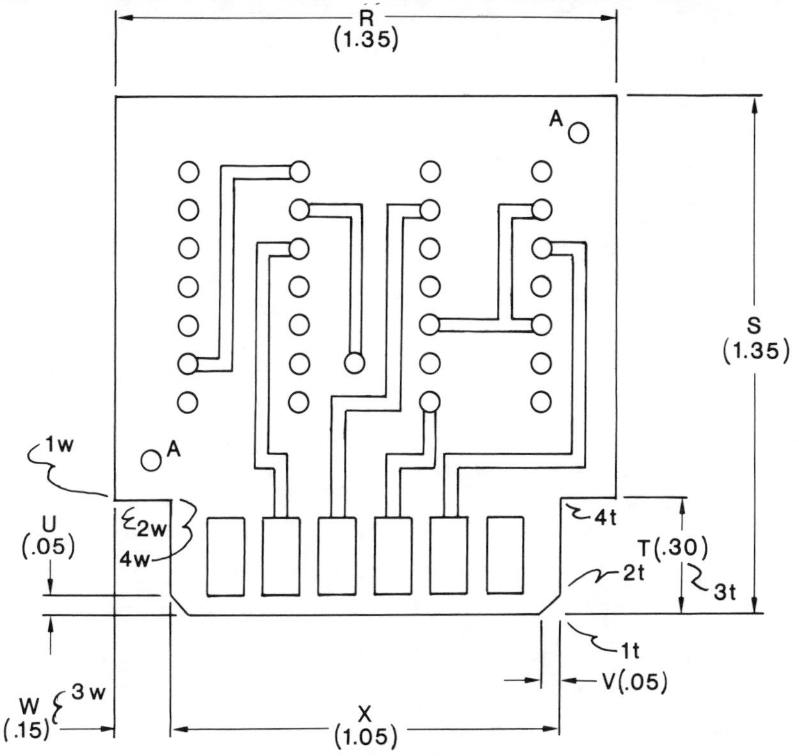

Fig. 1.
(#PCB012)

MAT'L: .062 THK EPOXY GLASS CLOTH LAMINATE WITH .0027 CU BOTH SIDES.
FINISH: GOLD FLASH ON ETCHED LETTERING AND CONDUCTORS BOTH SIDES.

P5 P6

P1

HOLE CHART			
DESCRIPTION	LETTER	DIA.	NO. REQ'D
LAND HOLES	UNMARKED	.032 AFTER PLATING THRU	29
TOOL HOLES	A	.062 NO PLATING THRU	2

P2

TITLE I.C. BOARD, CAD-DRILL AND CONTOUR					DWG. NO. CAD-21	
NAME	DATE	COURSE	GRADE	SCALE 2 X SIZE	SHEET OF	PAGE 104

IC BOARD CAD-ASSEMBLY: The PCB at right was originally modified from the drawing on page 102 as was PCB011.DWG. The artwork aids and IC hole patterns were CAD-erased, leaving only the board outline and six tabs. This modified drawing was filed as "PCB013.DWG" before the ICs, "ballooned 2," and tool holes were added. Program this CAD-assembly drawing by first recalling PCB013.DWG.

Exercise 1. Recalling the required drawing involves task—

☐ ______

1. Press R/R
2. ______
3. ______

Exercise 2. The second step involves recalling the ICs to points a and b. This requires task—

☐ ______

1. Press R/R
2. ______
3. ______
4. JS to point a (6, 27)
5. ______
6. ______
7. JS to point b (______)
8. Press U
9. Press 4

Exercise 3. The third step involves the CAD-drawing of "balloon 2." This could be done with a .40 dia. circle centered at point 2, or task—

☐ ______

1. R/R
2. ______
3. ______
4. ______

Exercise 4. While the circle task/routine is operative would be a good time to add the .062 dia. tool holes (.062 dia. drawn 2 × size = .125 dia., with a .06 radius) at points y and z.

5. JS to point y
6. ______
7. ______
8. ______

The balloon 2 leader was made using task—

☐ ______

The list of materials table (without the entries shown) was called up from a box "L" of a FORMS1 menu (task 34—similar to task 18) with its "h" in the lower right corner. The entries were made using task

☐ ______

(#PCB014)

ABBREVIATION USED:

ECN = Engineering change notice.

ITEM	NO. REQD.	REFERENCE DESIGNATION	DESCRIPTION	MANUFACTURER AND PART NUMBER OR MIL—TYPE DESIGNATION
3	2	U1, U2	INTEGRATED CIRCUIT	14 PIN I.C. F74921
2	1	PCB012	DRILLED BOARD	SHEET 2 OF 3 (PAGE 105)
1	X	———	SCHEMATIC	CAD-10; ECN #9.3

LIST OF MATERIAL h

TITLE IC BOARD CAD-ASSEMBLY					DWG. NO. CAD-22	
NAME	DATE	COURSE	GRADE	SCALE 2 X SIZE	SHEET OF	PAGE 105

Puck Modification: A CAD equipment company decided to use CAD-CAM to manufacture its puck housing (CAD-2, page 85). The CAD-CAM department needed to know where the buttons were located, and especially where the tangent point "a" (see Fig. 1 or Fig. 2) was located *by coordinates* relative to the cross-hair hole center origin point O(0,0). Since the puck is symmetrical, the right side can be mirrored from the left side. Therefore, only left side critical points are needed. The rear view of the puck was drawn for work on the rear door, but was also used to describe the location of "a" and button coordinates. Complete the started task and routine below.

Exercise 1. To CAD-draw the rectangular hidden buttons on the left side of Fig. 1 through bc and de requires task—

☐ ______________

1. R/R
2. ______________
3. Press ( ) at b, JS and
4.
5. Press ( ) at d, JS and
6.

Exercise 2. From the dimensions of the puck (page 85) and the origin at O(0,0) in Fig. 1, determine the exact coordinates of the following points: b(.55,.80), c( ), d( ), and e( ). [**Hint:** $x_b = .10/2 + .50 = .55$; while $y_b = 3.50 - 2.30 - .60 + .20 = .80$; actually $y_b = -.80$ but the ± signs are tentatively being omitted.] Insert the above coordinate entries in Fig. 1.

Exercise 3. To locate the tangent point "a", between the top arc and line af (Fig. 1), your supervisor developed part of the solution which he has asked you to complete. Use your portable calculator to complete the solution below—which normally would be calculated by the CAD system.

1. Line fo was drawn forming ∠s α and δ (Fig. 2).
2. Since fg and go are known, f0 and α can be determined (fg=.80; go=.60):

$$\tan \alpha = \frac{g0}{fg} = \frac{.60}{.80} = .75$$

α = ______

3. $\sin \alpha = \frac{.60}{fo} = (\quad)$
4. $fo = \frac{.60}{(\quad)}$; 5. f0 = ( )
6. $\sin \delta = \frac{.60}{f0} = \frac{.60}{(\quad)} =$
7. δ = ______
8. $\emptyset = 90 - \delta$
   $\emptyset = 90 -$
9. $\emptyset$ =
10. $\theta = \emptyset - \alpha$
    = ( )–( )
11. θ = ______
12. $\cos \theta = \frac{Xa}{.60}$

$\frac{Xa}{.60} = \cos(\quad) = (\_\_\_\_)$

13. Xa = .60( )
    Xa = ______
14. $\sin \theta = \frac{Ya}{.60}$

$\frac{Ya}{.60}$ = ______

Ya = ______

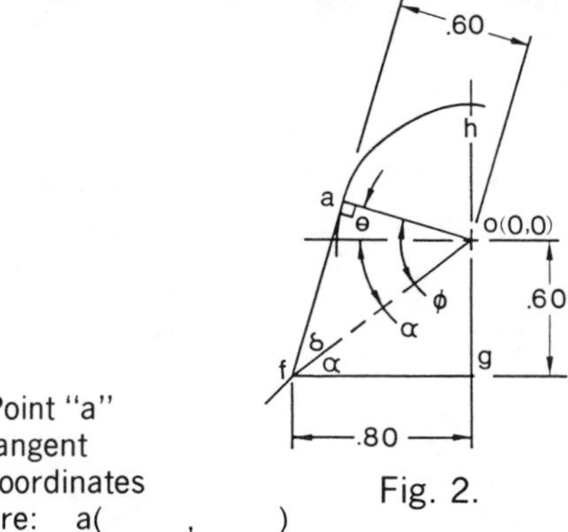

Fig. 1.

Fig. 2.

Point "a" tangent coordinates are: a(,)

Exercise 4. Complete the started task and routine below: To CAD-draw the arc ah with center "O" requires task—

☐ ______________

1. R/R
2. Type
3. JS to O Prs ( )
4.
5. JS to "a"

Exercise 5. The design department decided to widen the puck to allow for more button control functions. The new width was 2.50, making f'9' 1.05; but keeping all other dims. the same (see Fig. 3). Complete the new solution below to determine the coordinates of the new tangent point a'.

1. Line f'0' formed ∠s α' and δ': The student should draw f'0' and label α', δ', etc.
2. f'g' = 1.05; g'0' = .60

$$\tan \alpha' = \frac{g'o'}{f'g'} = \frac{.60}{1.05} = .571$$

α' = ______

3. $\sin \alpha' = \frac{.60}{f'o'} = (\_\_\_\_)$
4. $f'o' = \frac{.60}{(\quad)}$; 5. f'o' = (____)
6. $\sin \delta'$ = ______ =
7. δ' =
8. $\emptyset'$ = ______ = ______
9. $\emptyset'$ = ______
10. θ' = ______
    = ______
11. θ' = ______
12. $\cos \theta'$ = —— = (____)

$\frac{Xa'}{.60} = \cos(\quad) = (\_\_\_\_)$

13. Xa' = = ______
14. ______

Ya' = ______

Point "a'" tangent coordinates are: a' ( ). Complete the exercise below:

Exercise 6. To CAD-draw the arc a'h' with center O' requires task—

☐ ______________

1. ______________
2. ______________
3. ______________
4. ______________
5. ______________

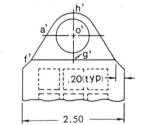

Fig. 3.

TITLE CAD-CAM PUCK MODIFICATION					DWG. NO. CAD-23	
NAME	DATE	COURSE	GRADE	SCALE	SHEET OF	PAGE 106

COMPONENT	SYMBOL	REF. DESIG.	APPENDIX B PAGE NO.
AMPLIFIER Two inputs		AR	
ANTENNA General Dipole		E	
BATTERY One cell Multicell	+ –	BT	
CAPACITOR General		C	112
Polarized (electrolytic)	+	C	113, 121
Variable or Adjustable		C	
CONDUCTOR Wire (general) Five-conductor cable		W	

COMPONENT	SYMBOL	REF. DESIG.	APPENDIX B PAGE NO.
Crossing not connected Junction (avoid if possible. Use next symbol) Junction		W	
CONNECTOR Male contact Female contact			
Separable connectors (engaged)	RECEPTACLE (J) OR PLUG	P J	
Engaged 4-conductor connectors; the plug has 1 male and 3 female contacts with individual contact designations shown in the complete-symbol column	A B C D OR (P) (J) A B C D	P J P J	

COMPONENT	SYMBOL	REF. DESIG.	APPENDIX B PAGE NO.
CONNECTOR ENGAGED COAXIAL Coaxial with the outside conductor shown carried through	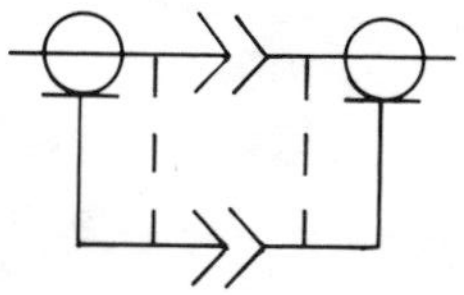		
Two-conductor (Jack)	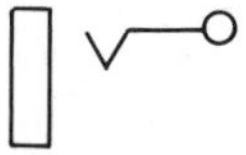	J	121
Two-conductor (Plug)	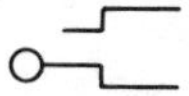	P	
Power, female contacts (2 conductors)	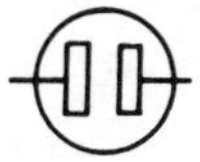		
Power, male contacts	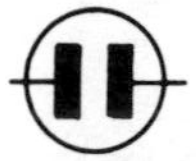		
CRYSTAL Piezoelectric	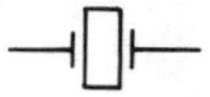	Y	120
FUSE	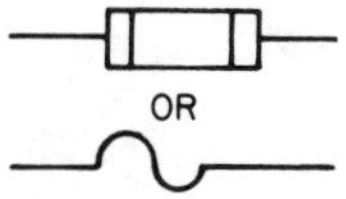 OR	F	120
GROUND Wire is bonded to chassis.	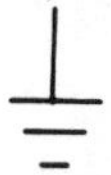		
Chassis or frame connection.	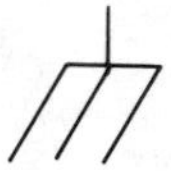		

COMPONENT	SYMBOL	REF. DESIG.	APPENDIX B PAGE NO.
INDUCTOR			
Magnetic core			
Tapped	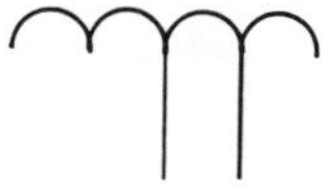	L	
Adjustable	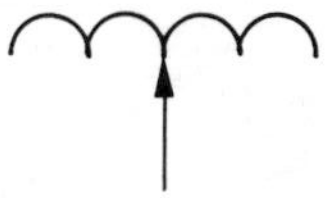		
LAMP			
AC type, neon		DS	
Incandescent filament	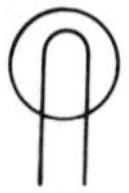		
MACHINE, ROTATING General	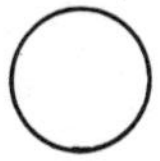		
Generator	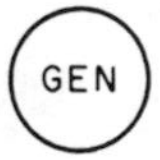 GEN	G	
Motor	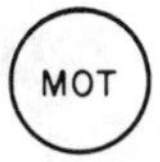 MOT	B	
MICROPHONE	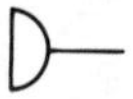	MK	

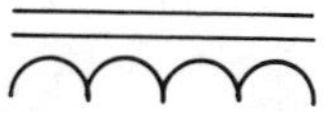

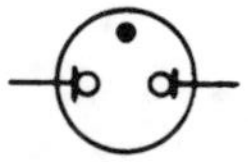

COMPONENT	SYMBOL	REF. DESIG.	APPENDIX B PAGE NO.
METER (INSTRUMENTS)			
General		M	
Example: milliammeter	MA		
RECTIFIER			
Semiconductor rectifier diode			114
Breakdown diode, unidirectional			114
		CR	
Breakdown diode, bidirectional			
Tunnel diode			114

COMPONENT	SYMBOL	REF. DESIG.	APPENDIX B PAGE NO.
RELAY			
Relay with transfer contacts		K	
	OR		
		K	
Polarized relay with transfer contacts	+ − +	K	

COMPONENT SYMBOLS

APPENDIX A

COMPONENT	SYMBOL	REF. DESIG.	APPENDIX B PAGE NO.	COMPONENT	SYMBOL	REF. DESIG.	APPENDIX B PAGE NO.
RESISTOR				SWITCH			
General	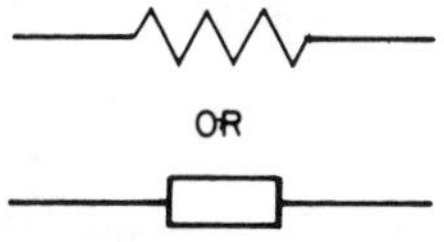 OR		115, 116	Single-Throw (ST)			
				Double-throw (DT)	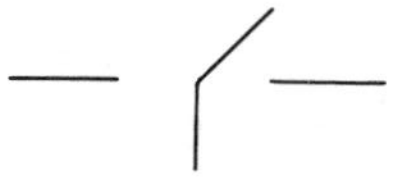		
Tapped	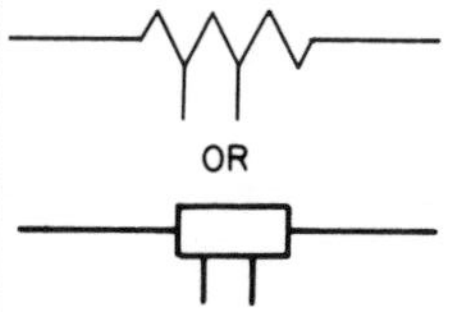 OR			Double-pole, Double-throw (DPDT) with terminals shown	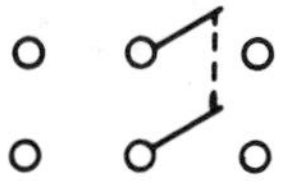		
				Push-button circuit closing (make)			
Adjustable	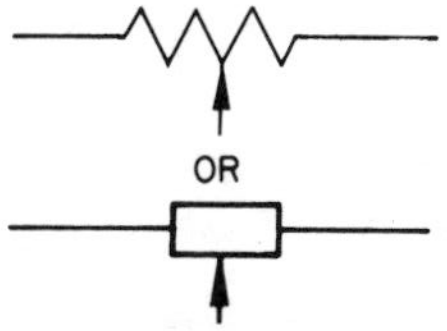 OR	R	117	Selector	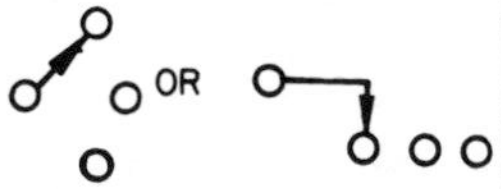 OR	S	117
Variable	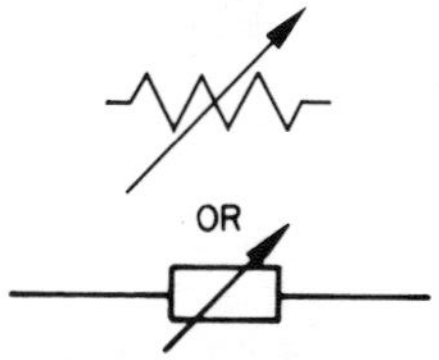 OR			Wafer, 3 poles, 3 circuits with 2 nonshorting and 1 shorting moving contacts.	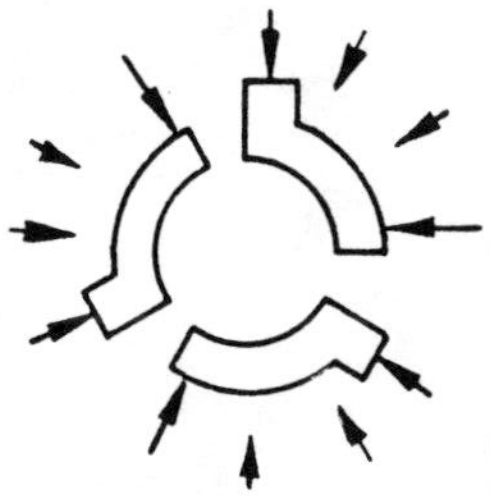		
SHIELDED CONDUCTOR				TERMINAL BOARD			
Single conductor				Group of 4 terminals	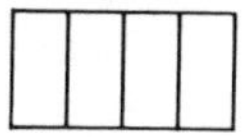	TB	
				TRANSFORMER			
Five-conductor cable, shield grounded	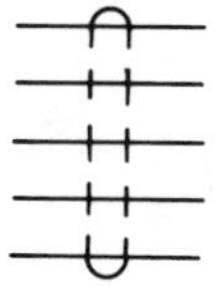	W		General	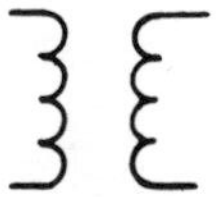		
				Magnetic core		T	120
SPEAKER							
General	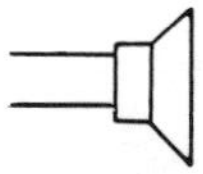	LS		Tapped	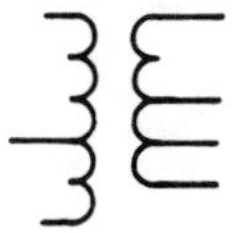		

COMPONENT SYMBOLS

APPENDIX A

COMPONENT	SYMBOL	REF. DESIG.	APPENDIX B PAGE NO.
TRANSISTOR			
PNP type			118
PNP type with 1 electrode connected to envelope	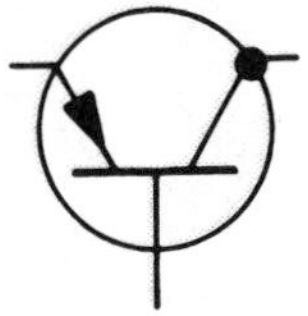		
NPN type	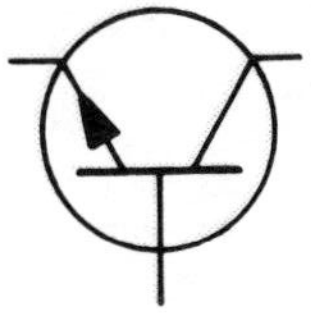	Q	
Unijunction N-type base	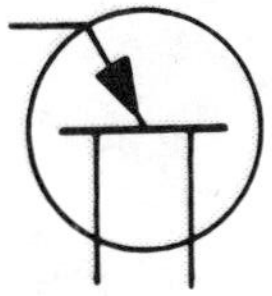		
Unijunction P-type base	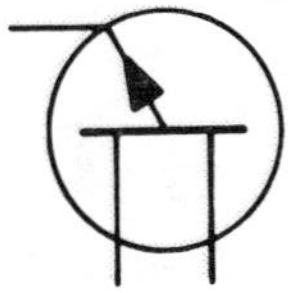		
Field-effect N-type base	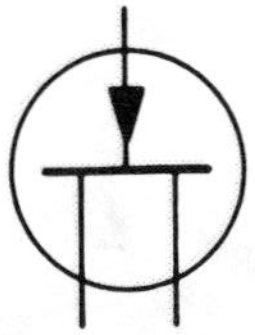		118

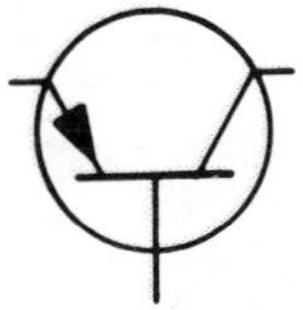

COMPONENT	SYMBOL	REF. DESIG.	APPENDIX B PAGE NO.
TUBE			
Triode	3, 5, 2, 7, 1, 8	V	
Twin triode	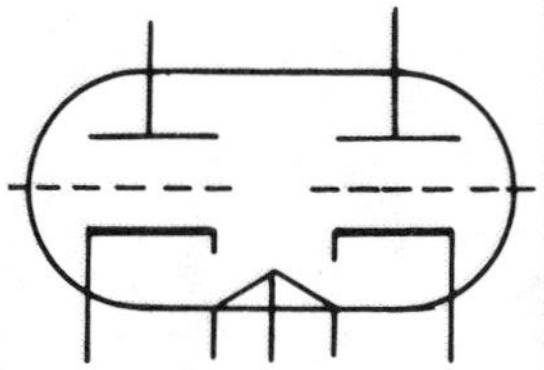		119
TUBE ENVELOPE			
General	OR		
Split			
Gas-filled	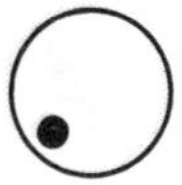		
C.R.T.	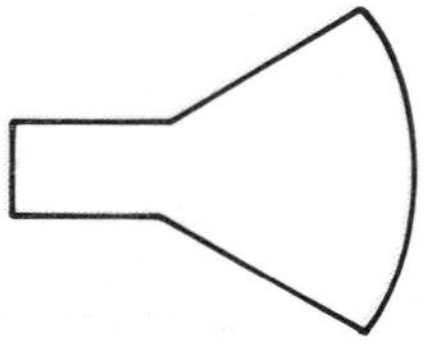		
WAVEGUIDE			
Circular		W	
Rectangular	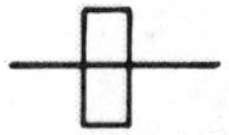		

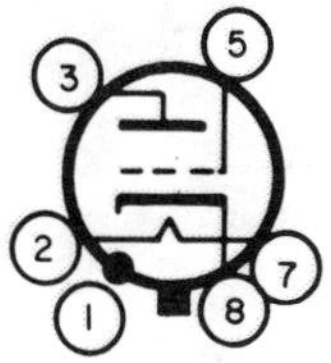

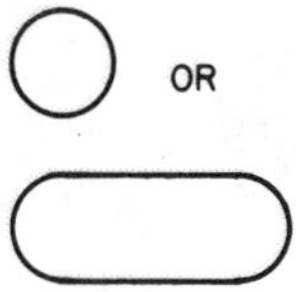

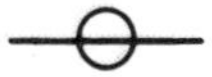

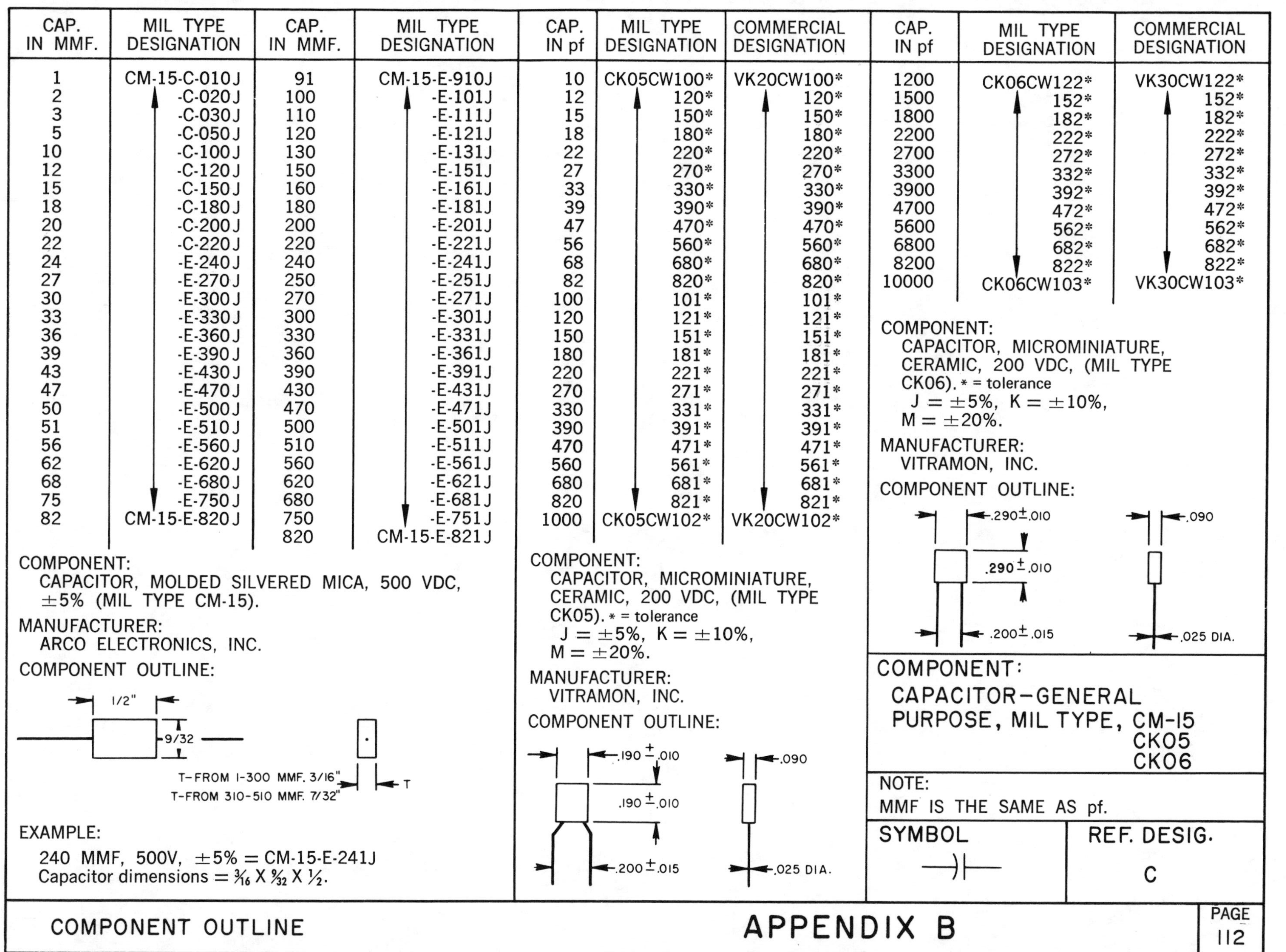

CAP. IN MMF.	MIL TYPE DESIGNATION
1	CM-15-C-010J
2	-C-020J
3	-C-030J
5	-C-050J
10	-C-100J
12	-C-120J
15	-C-150J
18	-C-180J
20	-C-200J
22	-C-220J
24	-E-240J
27	-E-270J
30	-E-300J
33	-E-330J
36	-E-360J
39	-E-390J
43	-E-430J
47	-E-470J
50	-E-500J
51	-E-510J
56	-E-560J
62	-E-620J
68	-E-680J
75	-E-750J
82	CM-15-E-820J

CAP. IN MMF.	MIL TYPE DESIGNATION
91	CM-15-E-910J
100	-E-101J
110	-E-111J
120	-E-121J
130	-E-131J
150	-E-151J
160	-E-161J
180	-E-181J
200	-E-201J
220	-E-221J
240	-E-241J
250	-E-251J
270	-E-271J
300	-E-301J
330	-E-331J
360	-E-361J
390	-E-391J
430	-E-431J
470	-E-471J
500	-E-501J
510	-E-511J
560	-E-561J
620	-E-621J
680	-E-681J
750	-E-751J
820	CM-15-E-821J

COMPONENT:
CAPACITOR, MOLDED SILVERED MICA, 500 VDC, ±5% (MIL TYPE CM-15).

MANUFACTURER:
ARCO ELECTRONICS, INC.

COMPONENT OUTLINE:

1/2"
9/32
T–FROM 1-300 MMF. 3/16"
T–FROM 310-510 MMF. 7/32"
T

EXAMPLE:

240 MMF, 500V, ±5% = CM-15-E-241J
Capacitor dimensions = 3/16 X 9/32 X 1/2.

CAP. IN pf	MIL TYPE DESIGNATION	COMMERCIAL DESIGNATION
10	CK05CW100*	VK20CW100*
12	120*	120*
15	150*	150*
18	180*	180*
22	220*	220*
27	270*	270*
33	330*	330*
39	390*	390*
47	470*	470*
56	560*	560*
68	680*	680*
82	820*	820*
100	101*	101*
120	121*	121*
150	151*	151*
180	181*	181*
220	221*	221*
270	271*	271*
330	331*	331*
390	391*	391*
470	471*	471*
560	561*	561*
680	681*	681*
820	821*	821*
1000	CK05CW102*	VK20CW102*

COMPONENT:
CAPACITOR, MICROMINIATURE, CERAMIC, 200 VDC, (MIL TYPE CK05). * = tolerance
J = ±5%, K = ±10%,
M = ±20%.

MANUFACTURER:
VITRAMON, INC.

COMPONENT OUTLINE:

.190 ± .010
.190 ± .010
.200 ± .015
.090
.025 DIA.

CAP. IN pf	MIL TYPE DESIGNATION	COMMERCIAL DESIGNATION
1200	CK06CW122*	VK30CW122*
1500	152*	152*
1800	182*	182*
2200	222*	222*
2700	272*	272*
3300	332*	332*
3900	392*	392*
4700	472*	472*
5600	562*	562*
6800	682*	682*
8200	822*	822*
10000	CK06CW103*	VK30CW103*

COMPONENT:
CAPACITOR, MICROMINIATURE, CERAMIC, 200 VDC, (MIL TYPE CK06). * = tolerance
J = ±5%, K = ±10%,
M = ±20%.

MANUFACTURER:
VITRAMON, INC.

COMPONENT OUTLINE:

.290 ± .010
.290 ± .010
.200 ± .015
.090
.025 DIA.

COMPONENT:
CAPACITOR–GENERAL PURPOSE, MIL TYPE, CM-15
CK05
CK06

NOTE:
MMF IS THE SAME AS pf.

SYMBOL	REF. DESIG.
—)|—	C

COMPONENT OUTLINE

APPENDIX B

CAPACITANCE IN MFD.	DC RATED VOLTAGE	CASE SIZE	MIL TYPE DESIGNATION
5.6	6	A	CS13AB5R6K
6.8	6	A	AB6R8K
47	6	B	AB470K
56	6	B	AB560K
150	6	C	AB151K
180	6	C	AB181K
270	6	D	AB271K
330	6	D	AB331K
3.9	10	A	AC3R9K
4.7	10	A	AC4R7K
27	10	B	AC270K
33	10	B	AC330K
39	10	B	AC390K
82	10	C	AC820K
100	10	C	AC101K
120	10	C	AC121K
180	10	D	AC181K
220	10	D	AC221K
2.7	15	A	AD2R7K
3.3	15	A	AD3R3K
18	15	B	AD180K
22	15	B	AD220K
56	15	C	AD560K
68	15	C	AD680K
120	15	D	AD121K
150	15	D	AD151K
1.2	20	A	AE1R2K
1.5	20	A	CS13AE1R5K

CAPACITANCE IN MFD.	DC RATED VOLTAGE	CASE SIZE	MIL TYPE DESIGNATION
1.8	20	A	CS13AE1R8K
2.2	20	A	AE2R2K
8.2	20	B	AE8R2K
10	20	B	AE100K
12	20	B	AE120K
15	20	B	AE150K
27	20	C	AE270K
33	20	C	AE330K
39	20	C	AE390K
47	20	C	AE470K
56	20	D	AE560K
68	20	D	AE680K
82	20	D	AE820K
100	20	D	AE101K
0.33	35	A	AFR33K
0.39	35	A	AFR39K
0.47	35	A	AFR47K
0.56	35	A	AFR56K
0.68	35	A	AFR68K
0.82	35	A	AFR82K
1	35	A	AF010K
1.2	35	B	AF1R2K
1.5	35	B	AF1R5K
1.8	35	B	AF1R8K
2.2	35	B	AF2R2K
2.7	35	B	AF2R7K
3.3	35	B	AF3R3K
3.9	35	B	AF3R9K
4.7	35	B	AF4R7K
5.6	35	B	CS13AF5R6K

CAPACITANCE IN MFD.	DC RATED VOLTAGE	CASE SIZE	MIL TYPE DESIGNATION
6.8	35	B	CS13AF6R8K
8.2	35	C	AF8R2K
10	35	C	AF100K
12	35	C	AF120K
15	35	C	AF150K
18	35	C	AF180K
22	35	C	AF220K
27	35	D	AF270K
33	35	D	AF330K
39	35	D	AF390K
47	35	D	AF470K
1	50	A	AG010K
1.2	50	B	AG1R2K
1.5	50	B	AG1R5K
1.8	50	B	AG1R8K
2.2	50	B	AG2R2K
2.7	50	B	AG2R7K
3.3	50	B	AG3R3K
3.9	50	B	AG3R9K
4.7	50	B	AG4R7K
5.6	50	C	AG5R6K
6.8	50	C	AG6R8K
8.2	50	C	AG8R2K
10	50	C	AG100K
12	50	C	AG120K
15	50	C	AG150K
18	50	C	AG180K
22	50	D	CS13AG220K

COMPONENT OUTLINE
(SEE DIMENSIONS)

L — C — $1\frac{1}{2} \pm \frac{1}{4}$ — D

EXAMPLE:

0.68 MFD.35V, ±10% = Mil No. CS13AFR68K
For Capacitor Dimensions see Case A

CAPACITOR DIMENSIONS

CASE SIZE	C max	D 0+.016 −0.015	L±0.031	LEAD DIAMETER +0.005 −0.001
A	0.422	0.135	0.286	0.020
B	0.610	0.185	0.474	0.020
C	0.822	0.289	0.686	0.025
D	0.922	0.351	0.786	0.025

SYMBOL

COMPONENT
CAPACITOR, FIXED, SOLID ELECTROLYTE, TANTALUM, MIL TYPE CS13 (±10%)

REF. DESIG.
C

APPENDIX B

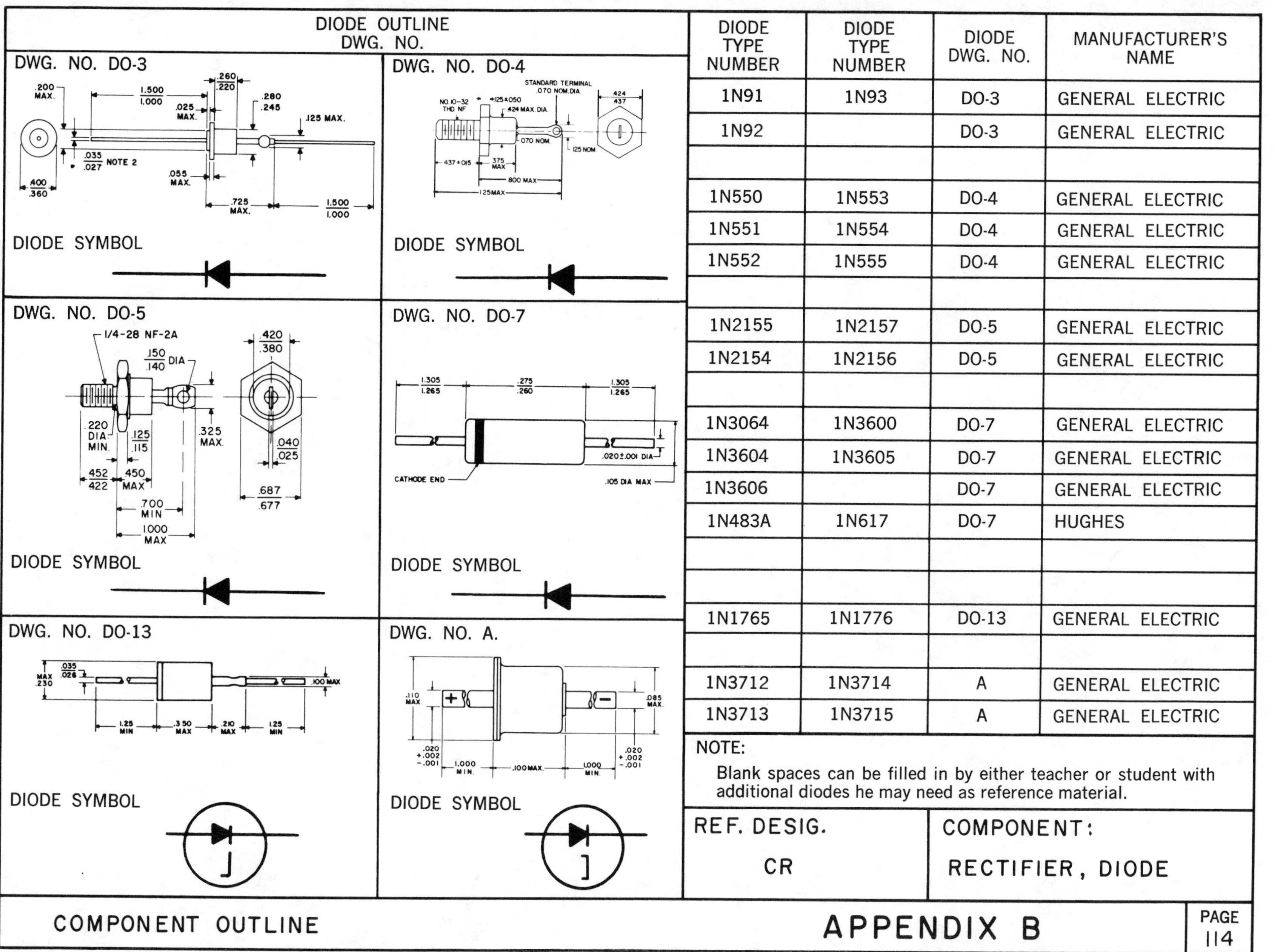

DIODE TYPE NUMBER	DIODE TYPE NUMBER	DIODE DWG. NO.	MANUFACTURER'S NAME
1N91	1N93	DO-3	GENERAL ELECTRIC
1N92		DO-3	GENERAL ELECTRIC
1N550	1N553	DO-4	GENERAL ELECTRIC
1N551	1N554	DO-4	GENERAL ELECTRIC
1N552	1N555	DO-4	GENERAL ELECTRIC
1N2155	1N2157	DO-5	GENERAL ELECTRIC
1N2154	1N2156	DO-5	GENERAL ELECTRIC
1N3064	1N3600	DO-7	GENERAL ELECTRIC
1N3604	1N3605	DO-7	GENERAL ELECTRIC
1N3606		DO-7	GENERAL ELECTRIC
1N483A	1N617	DO-7	HUGHES
1N1765	1N1776	DO-13	GENERAL ELECTRIC
1N3712	1N3714	A	GENERAL ELECTRIC
1N3713	1N3715	A	GENERAL ELECTRIC

NOTE:

Blank spaces can be filled in by either teacher or student with additional diodes he may need as reference material.

REF. DESIG.: CR

COMPONENT: RECTIFIER, DIODE

COMPONENT OUTLINE

APPENDIX B

Total Resistance Ohms	MIL Type Designation	Total Resistance Ohms	MIL Type Designation	Total Resistance Ohms	MIL Type Designation	Total Resistance Ohms	MIL Type Designation	Total Resistance Megohms	MIL Type Designation
10	RC07GF100J	180	RC07GF181J	3,300	RC07GF332J	56,000	RC07GF563J	1.0	RC07GF105J
11	110J	200	201J	3,600	362J	62,000	623J	1.1	115J
12	120J	220	221J	3,900	392J	68,000	683J	1.2	125J
13	130J	240	241J	4,300	432J	75,000	753J	1.3	135J
15	150J	270	271J	4,700	472J	82,000	823J	1.5	155J
16	160J	300	301J	5,100	512J	91,000	913J	1.6	165J
18	180J	330	331J	5,600	562J	100,000	104J	1.8	185J
20	200J	360	361J	6,200	622J	110,000	114J	2.0	205J
22	220J	390	391J	6,800	682J	120,000	124J	2.2	225J
24	240J	430	431J	7,500	752J	130,000	134J	2.4	245J
								2.7	275J
27	270J	470	471J	8,200	822J	150,000	154J	3.0	305J
30	300J	510	511J	9,100	912J	160,000	164J	3.3	335J
33	330J	560	561J	10,000	103J	180,000	184J	3.6	365J
36	360J	620	621J	11,000	113J	200,000	204J	3.9	395J
39	390J	680	681J	12,000	123J	220,000	224J	4.3	435J
43	430J	750	751J	13,000	133J	240,000	244J	4.7	475J
47	470J	820	821J	15,000	153J	270,000	274J	5.1	515J
51	510J	910	911J	16,000	163J	300,000	304J	5.6	565J
56	560J	1,000	102J	18,000	183J	330,000	334J	6.2	625J
62	620J	1,100	112J	20,000	203J	360,000	364J	6.8	685J
								7.5	755J
68	680J	1,200	122J	22,000	223J	390,000	394J	8.2	825J
75	750J	1,300	132J	24,000	243J	430,000	434J	9.1	915J
82	820J	1,500	152J	27,000	273J	470,000	474J	10.0	106J
91	910J	1,600	162J	30,000	303J	510,000	514J	11	116J
100	101J	1,800	182J	33,000	333J	560,000	564J	12	126J
110	111J	2,000	202J	36,000	363J	620,000	624J	13	136J
120	121J	2,200	222J	39,000	393J	680,000	684J	15	156J
130	131J	2,400	242J	43,000	433J	750,000	754J	16	166J
150	151J	2,700	272J	47,000	473J	820,000	824J	18	186J
160	RC07GF161J	3,000	RC07GF302J	51,000	RC07GF513J	910,000	RC07GF914J	20	206J
								22	RC07GF226J

COMPONENT OUTLINE

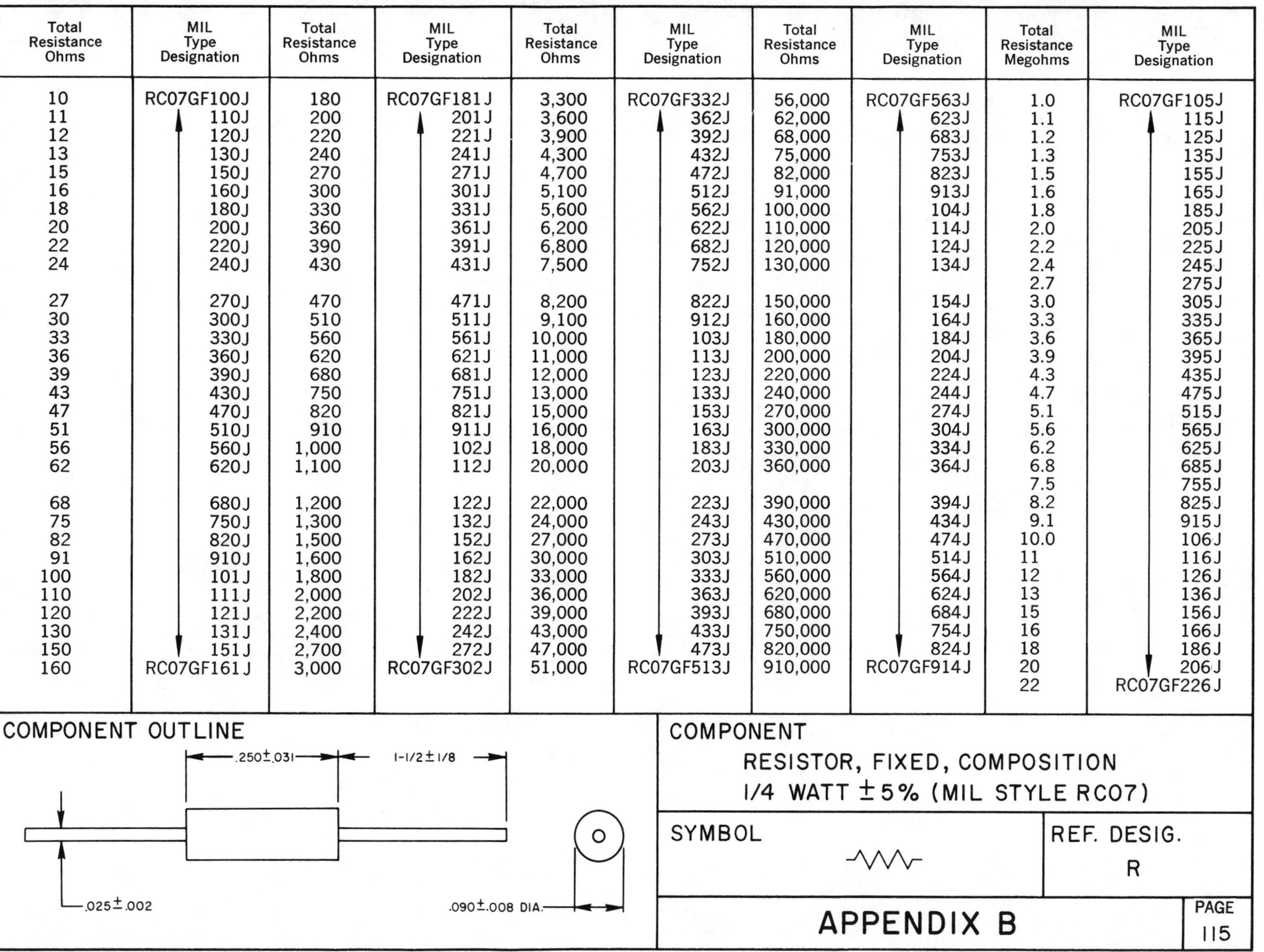

COMPONENT

RESISTOR, FIXED, COMPOSITION
1/4 WATT ±5% (MIL STYLE RC07)

SYMBOL

REF. DESIG.
R

APPENDIX B

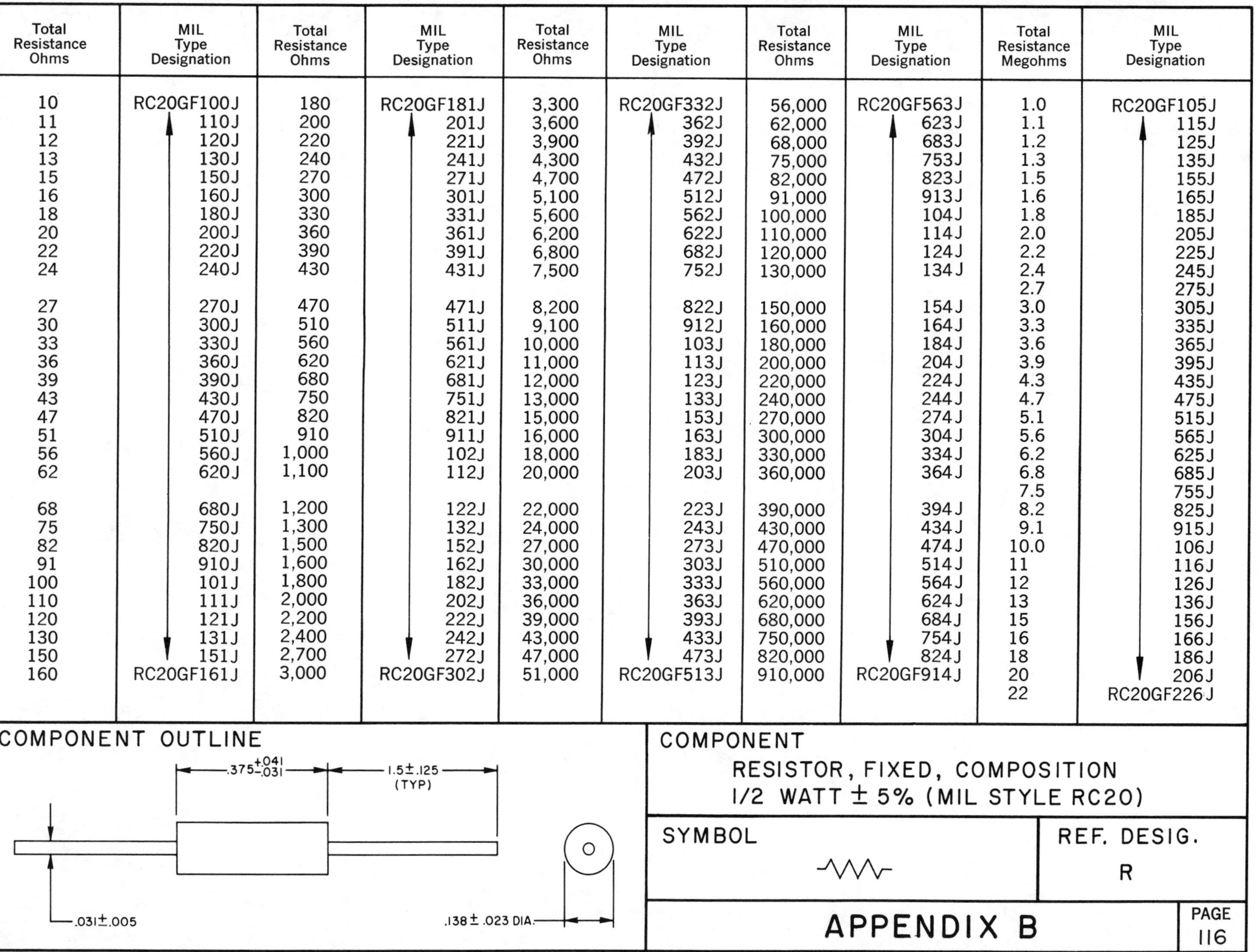

Total Resistance Ohms	MIL Type Designation	Total Resistance Ohms	MIL Type Designation	Total Resistance Ohms	MIL Type Designation	Total Resistance Ohms	MIL Type Designation	Total Resistance Megohms	MIL Type Designation
10	RC20GF100J	180	RC20GF181J	3,300	RC20GF332J	56,000	RC20GF563J	1.0	RC20GF105J
11	110J	200	201J	3,600	362J	62,000	623J	1.1	115J
12	120J	220	221J	3,900	392J	68,000	683J	1.2	125J
13	130J	240	241J	4,300	432J	75,000	753J	1.3	135J
15	150J	270	271J	4,700	472J	82,000	823J	1.5	155J
16	160J	300	301J	5,100	512J	91,000	913J	1.6	165J
18	180J	330	331J	5,600	562J	100,000	104J	1.8	185J
20	200J	360	361J	6,200	622J	110,000	114J	2.0	205J
22	220J	390	391J	6,800	682J	120,000	124J	2.2	225J
24	240J	430	431J	7,500	752J	130,000	134J	2.4	245J
								2.7	275J
27	270J	470	471J	8,200	822J	150,000	154J	3.0	305J
30	300J	510	511J	9,100	912J	160,000	164J	3.3	335J
33	330J	560	561J	10,000	103J	180,000	184J	3.6	365J
36	360J	620	621J	11,000	113J	200,000	204J	3.9	395J
39	390J	680	681J	12,000	123J	220,000	224J	4.3	435J
43	430J	750	751J	13,000	133J	240,000	244J	4.7	475J
47	470J	820	821J	15,000	153J	270,000	274J	5.1	515J
51	510J	910	911J	16,000	163J	300,000	304J	5.6	565J
56	560J	1,000	102J	18,000	183J	330,000	334J	6.2	625J
62	620J	1,100	112J	20,000	203J	360,000	364J	6.8	685J
								7.5	755J
68	680J	1,200	122J	22,000	223J	390,000	394J	8.2	825J
75	750J	1,300	132J	24,000	243J	430,000	434J	9.1	915J
82	820J	1,500	152J	27,000	273J	470,000	474J	10.0	106J
91	910J	1,600	162J	30,000	303J	510,000	514J	11	116J
100	101J	1,800	182J	33,000	333J	560,000	564J	12	126J
110	111J	2,000	202J	36,000	363J	620,000	624J	13	136J
120	121J	2,200	222J	39,000	393J	680,000	684J	15	156J
130	131J	2,400	242J	43,000	433J	750,000	754J	16	166J
150	151J	2,700	272J	47,000	473J	820,000	824J	18	186J
160	RC20GF161J	3,000	RC20GF302J	51,000	RC20GF513J	910,000	RC20GF914J	20	206J
								22	RC20GF226J

COMPONENT OUTLINE

COMPONENT

RESISTOR, FIXED, COMPOSITION
1/2 WATT ± 5% (MIL STYLE RC20)

SYMBOL

REF. DESIG.
R

APPENDIX B

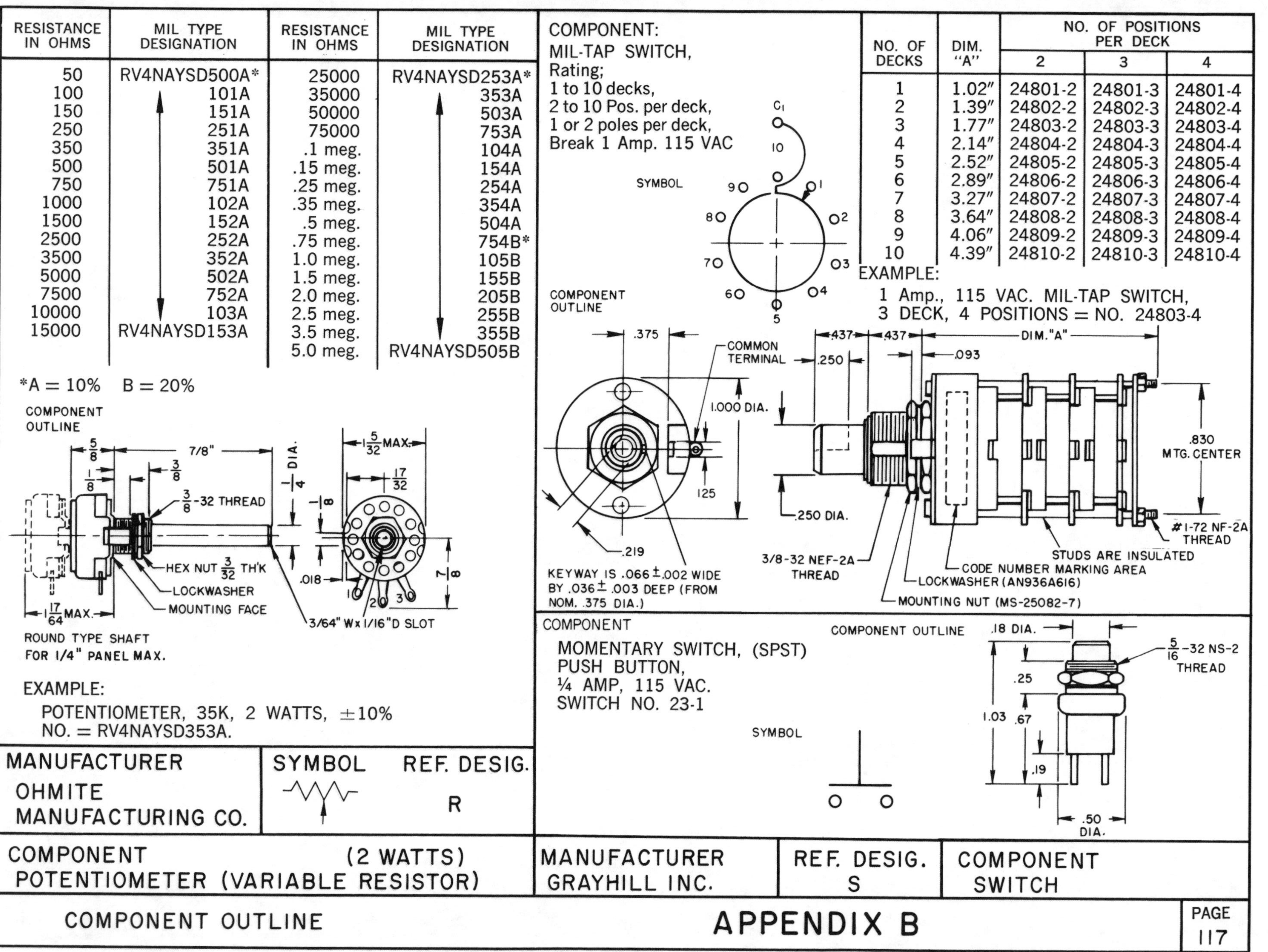

RESISTANCE IN OHMS	MIL TYPE DESIGNATION	RESISTANCE IN OHMS	MIL TYPE DESIGNATION
50	RV4NAYSD500A*	25000	RV4NAYSD253A*
100	101A	35000	353A
150	151A	50000	503A
250	251A	75000	753A
350	351A	.1 meg.	104A
500	501A	.15 meg.	154A
750	751A	.25 meg.	254A
1000	102A	.35 meg.	354A
1500	152A	.5 meg.	504A
2500	252A	.75 meg.	754B*
3500	352A	1.0 meg.	105B
5000	502A	1.5 meg.	155B
7500	752A	2.0 meg.	205B
10000	103A	2.5 meg.	255B
15000	RV4NAYSD153A	3.5 meg.	355B
		5.0 meg.	RV4NAYSD505B

*A = 10% B = 20%

EXAMPLE:

POTENTIOMETER, 35K, 2 WATTS, ±10%
NO. = RV4NAYSD353A.

MANUFACTURER	SYMBOL	REF. DESIG.
OHMITE MANUFACTURING CO.		R

COMPONENT (2 WATTS)
POTENTIOMETER (VARIABLE RESISTOR)

COMPONENT:
MIL-TAP SWITCH,
Rating;
1 to 10 decks,
2 to 10 Pos. per deck,
1 or 2 poles per deck,
Break 1 Amp. 115 VAC

NO. OF DECKS	DIM. "A"	NO. OF POSITIONS PER DECK 2	3	4
1	1.02"	24801-2	24801-3	24801-4
2	1.39"	24802-2	24802-3	24802-4
3	1.77"	24803-2	24803-3	24803-4
4	2.14"	24804-2	24804-3	24804-4
5	2.52"	24805-2	24805-3	24805-4
6	2.89"	24806-2	24806-3	24806-4
7	3.27"	24807-2	24807-3	24807-4
8	3.64"	24808-2	24808-3	24808-4
9	4.06"	24809-2	24809-3	24809-4
10	4.39"	24810-2	24810-3	24810-4

EXAMPLE:
1 Amp., 115 VAC. MIL-TAP SWITCH,
3 DECK, 4 POSITIONS = NO. 24803-4

COMPONENT

MOMENTARY SWITCH, (SPST)
PUSH BUTTON,
¼ AMP, 115 VAC.
SWITCH NO. 23-1

MANUFACTURER	REF. DESIG.	COMPONENT
GRAYHILL INC.	S	SWITCH

COMPONENT OUTLINE

APPENDIX B

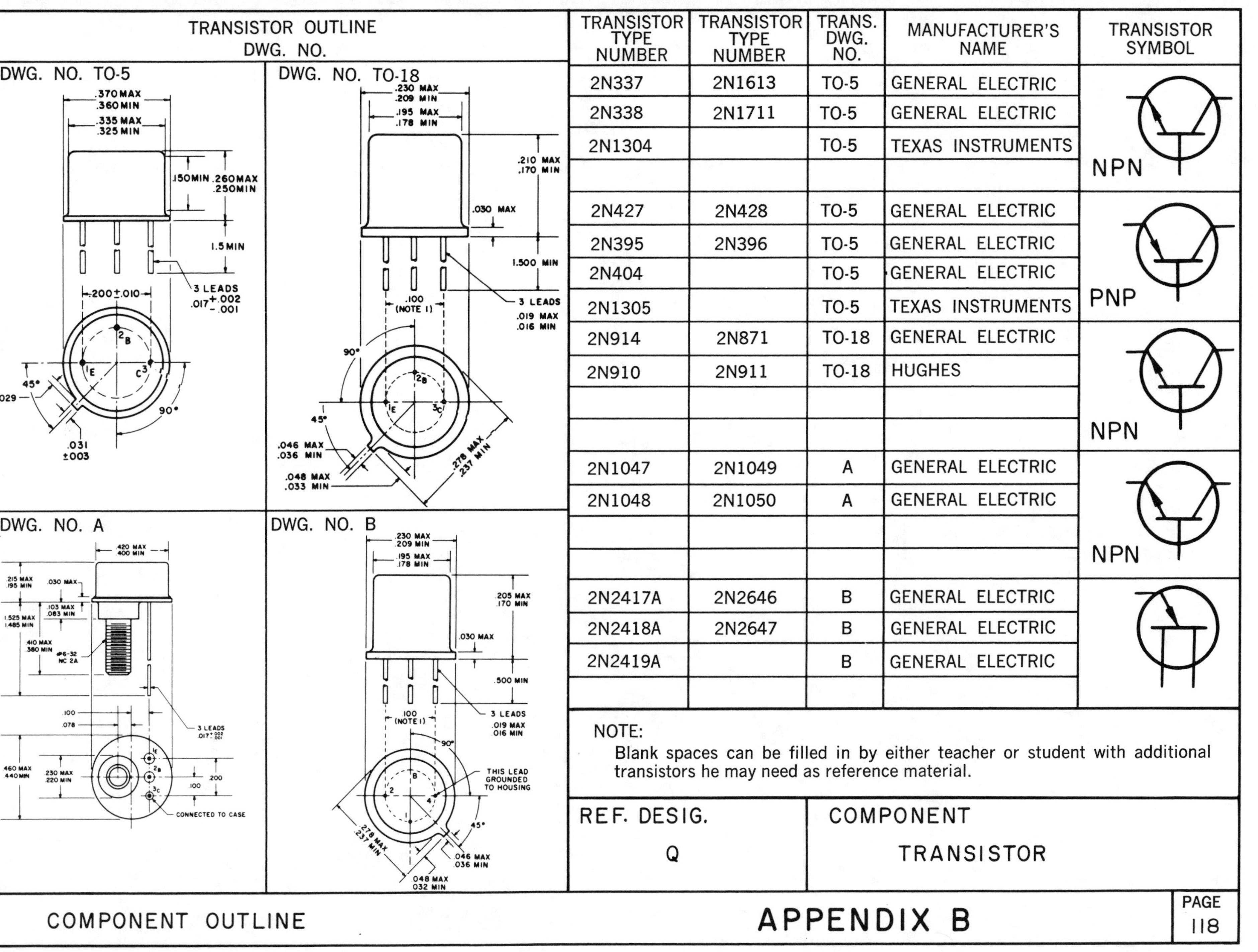

TRANSISTOR TYPE NUMBER	TRANSISTOR TYPE NUMBER	TRANS. DWG. NO.	MANUFACTURER'S NAME	TRANSISTOR SYMBOL
2N337	2N1613	TO-5	GENERAL ELECTRIC	NPN
2N338	2N1711	TO-5	GENERAL ELECTRIC	
2N1304		TO-5	TEXAS INSTRUMENTS	
2N427	2N428	TO-5	GENERAL ELECTRIC	PNP
2N395	2N396	TO-5	GENERAL ELECTRIC	
2N404		TO-5	GENERAL ELECTRIC	
2N1305		TO-5	TEXAS INSTRUMENTS	
2N914	2N871	TO-18	GENERAL ELECTRIC	NPN
2N910	2N911	TO-18	HUGHES	
2N1047	2N1049	A	GENERAL ELECTRIC	NPN
2N1048	2N1050	A	GENERAL ELECTRIC	
2N2417A	2N2646	B	GENERAL ELECTRIC	
2N2418A	2N2647	B	GENERAL ELECTRIC	
2N2419A		B	GENERAL ELECTRIC	

NOTE:
Blank spaces can be filled in by either teacher or student with additional transistors he may need as reference material.

REF. DESIG.	COMPONENT
Q	TRANSISTOR

COMPONENT OUTLINE

APPENDIX B

TUBE DIMENSIONS DWG. NO.

DWG. NO. 1 — Miniature Button 7-Pin Base

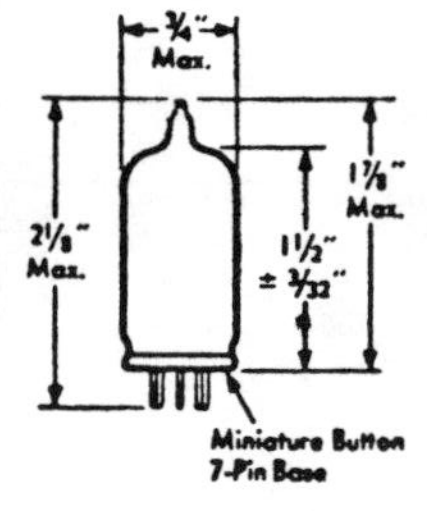

DWG. NO. 2 — Small Wafer Octal Base

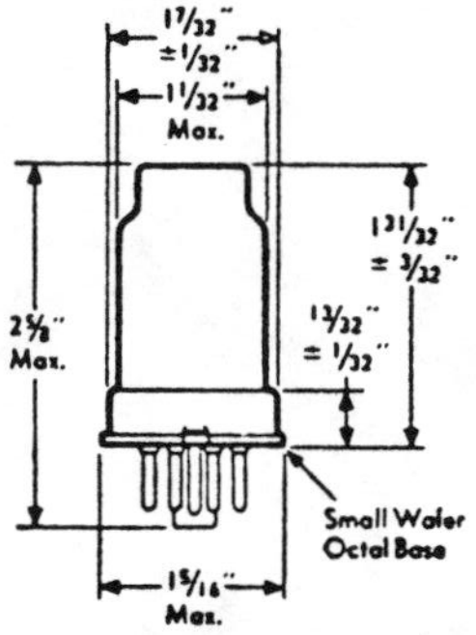

DWG. NO. 3 — Medium Shell Octal Base

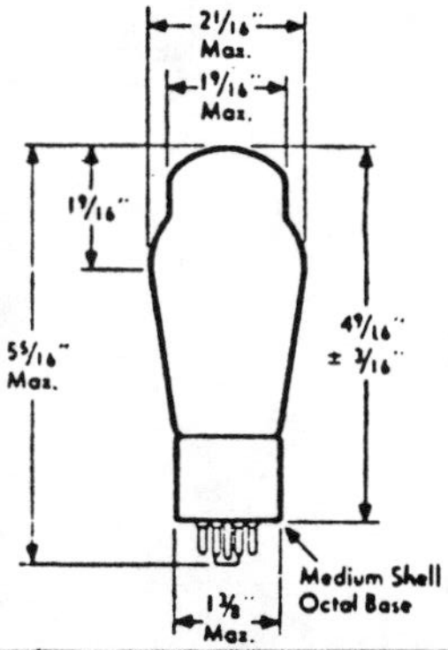

DWG. NO. 4 — Small-Button 9-Pin Base

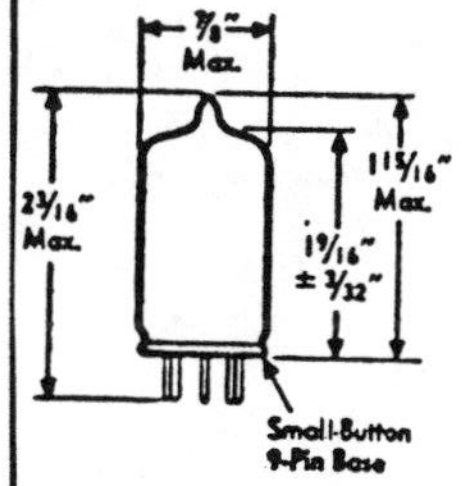

DWG. NO. 5 — Miniature Cap; Small Wafer Octal Base

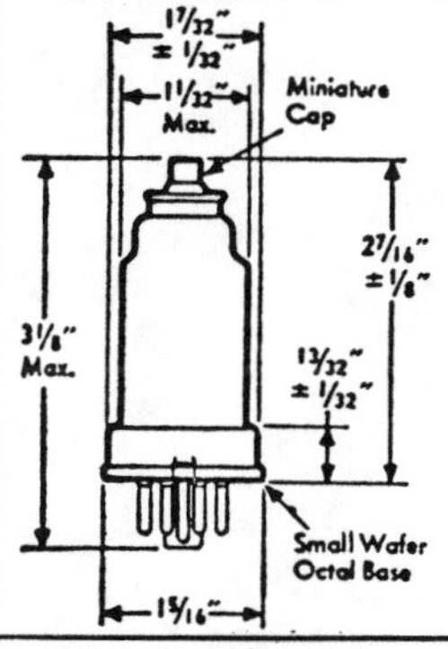

DWG. NO. 6 — Miniature Button 7-Pin Base

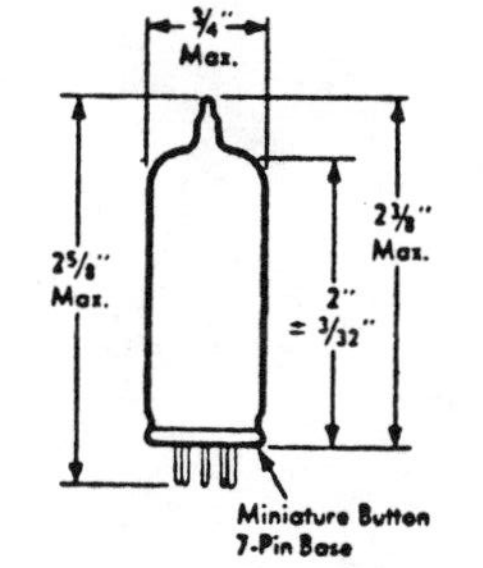

DWG. NO. 7 — Miniature Button 7-Pin Base

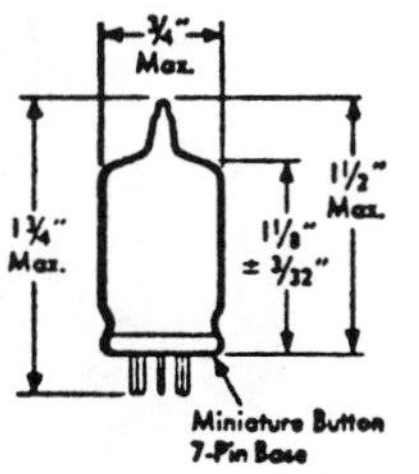

DWG. NO. 8 — Small Wafer Octal Base

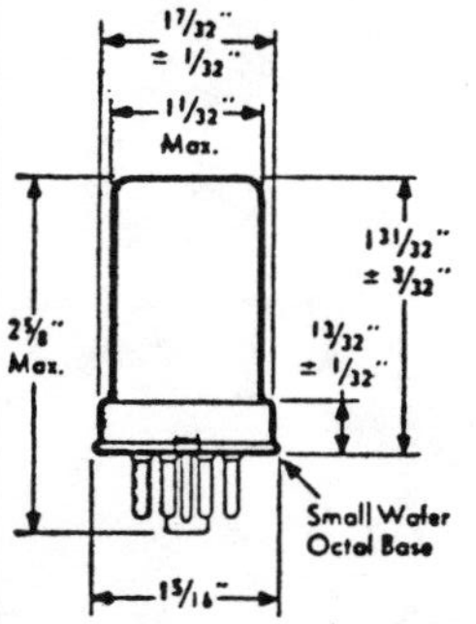

DWG. NO. 9 — Small Cap; Short Intermediate Shell Octal 6-Pin Base

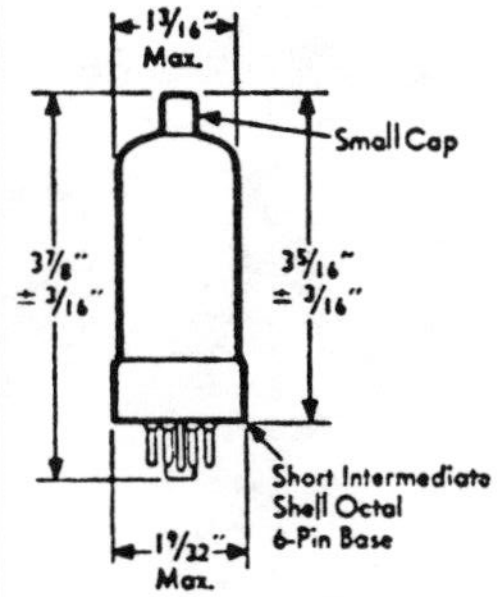

TUBE TYPE NUMBER	TUBE SYMBOL	TUBE DIMENSION DWG. NO.	TUBE TYPE NUMBER	TUBE SYMBOL	TUBE DIMENSION DWG. NO.
OZ4	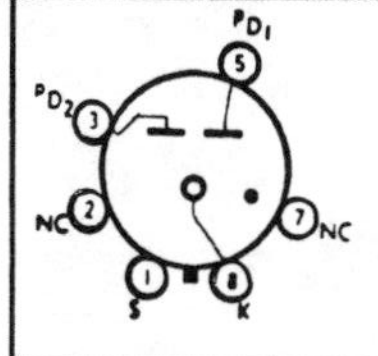	2	6AL5	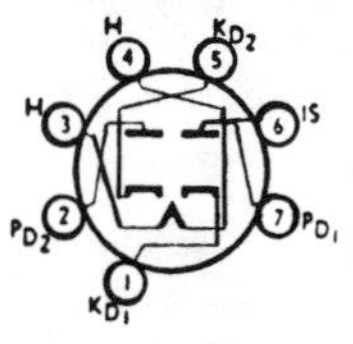	7
1B3-GT	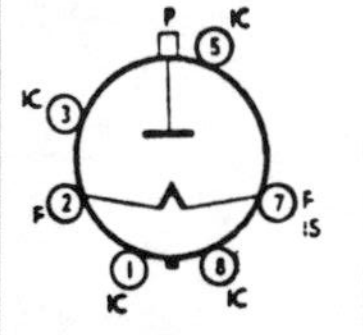	9	6AQ6 6AT6 6BF6 12AV6	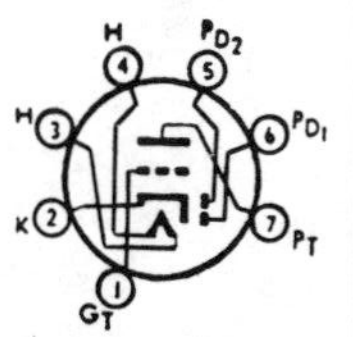	1
5R4-GY 5U4-G	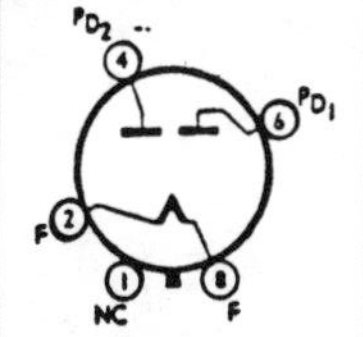	3	6BN6	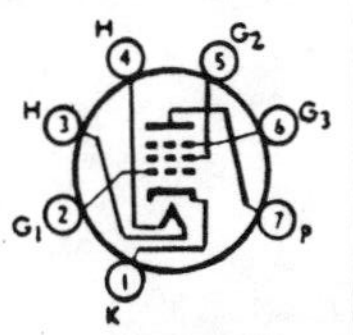	6
6AB7 6AC7 12SJ7	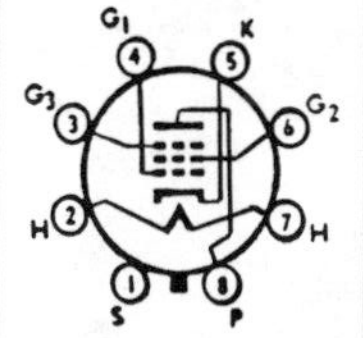	8	6Q7 6R7	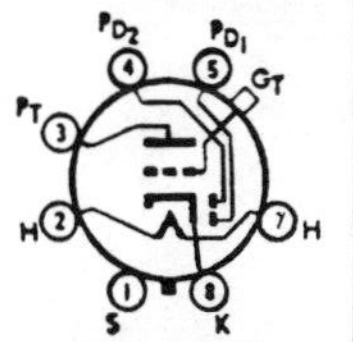	5
6AK5	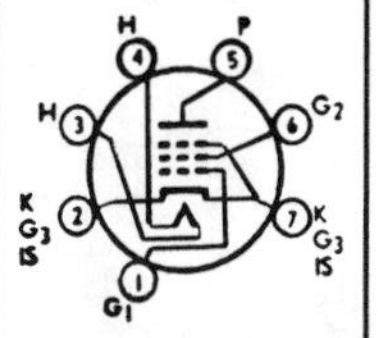	7	12AX7 12AY7	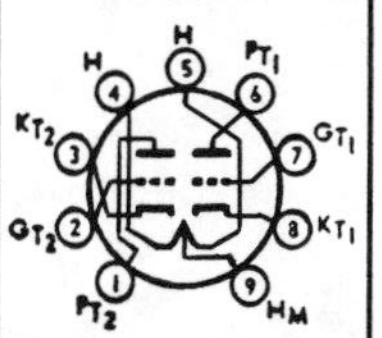	4
6AK6 6AU6 12BA6	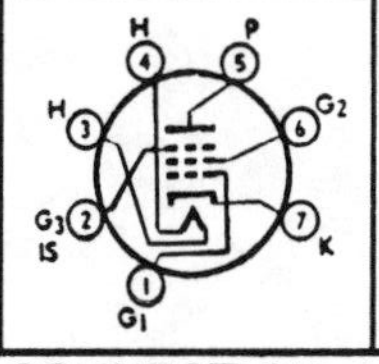	1	35W4	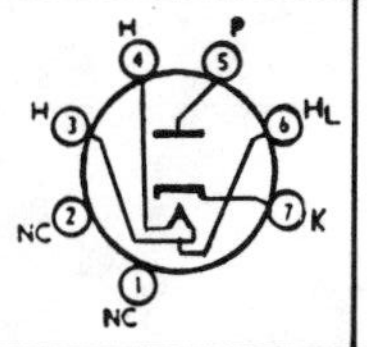	6

PIN CONNECTION MARKING

F — Filament
G — Grid
H — Heater
K — Cathode
NC — No connection
P — Plate or Anode

REF. DESIG.: V

COMPONENT: TUBE

MANUFACTURING: WESTINGHOUSE ELECTRIC CORP.

COMPONENT OUTLINE

APPENDIX B

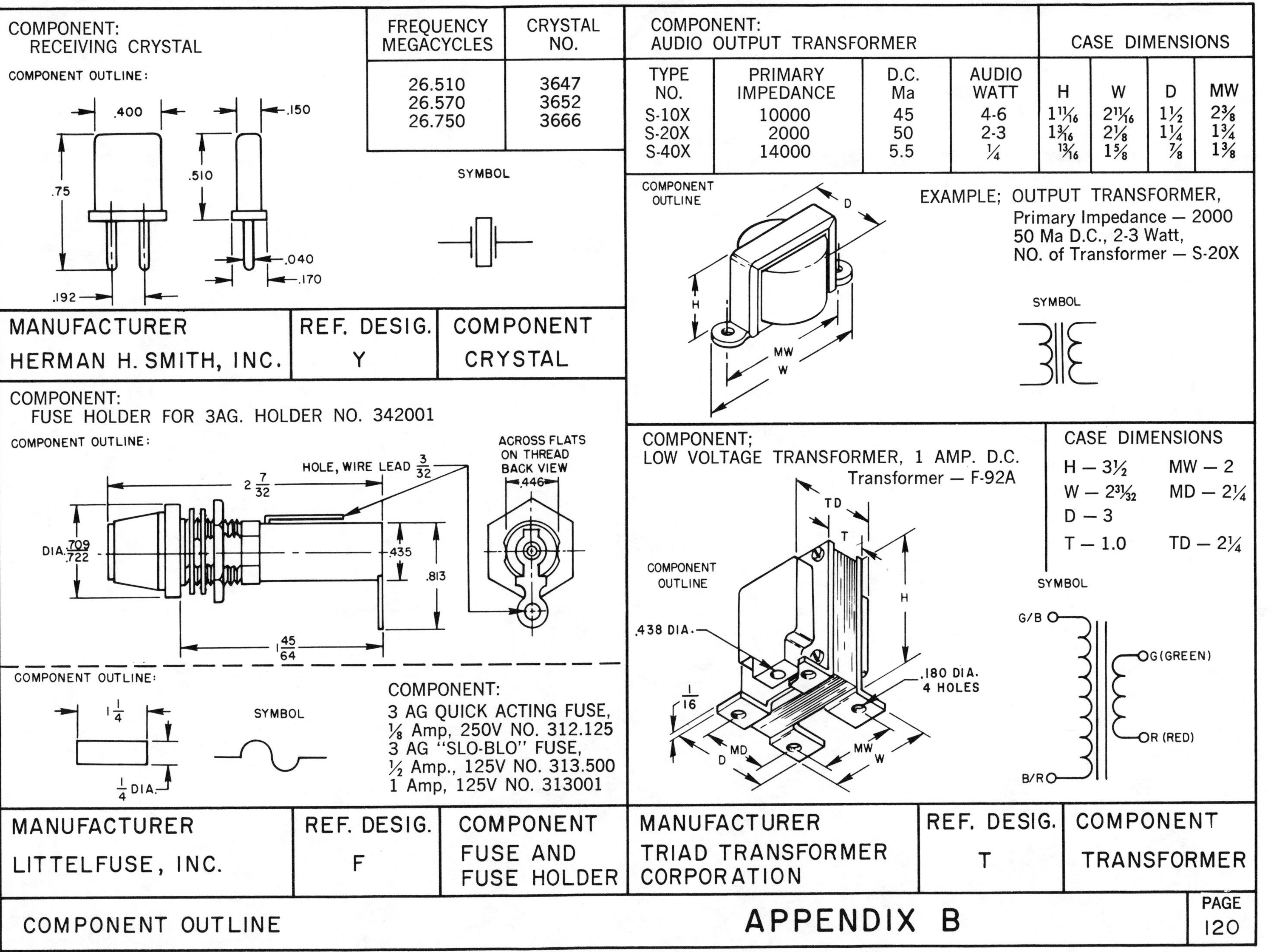

COMPONENT:
RECEIVING CRYSTAL

COMPONENT OUTLINE:

FREQUENCY MEGACYCLES	CRYSTAL NO.
26.510	3647
26.570	3652
26.750	3666

MANUFACTURER	REF. DESIG.	COMPONENT
HERMAN H. SMITH, INC.	Y	CRYSTAL

COMPONENT:
FUSE HOLDER FOR 3AG. HOLDER NO. 342001

COMPONENT OUTLINE:

COMPONENT OUTLINE:

COMPONENT:
3 AG QUICK ACTING FUSE,
⅛ Amp, 250V NO. 312.125
3 AG "SLO-BLO" FUSE,
½ Amp., 125V NO. 313.500
1 Amp, 125V NO. 313001

MANUFACTURER	REF. DESIG.	COMPONENT
LITTELFUSE, INC.	F	FUSE AND FUSE HOLDER

COMPONENT:
AUDIO OUTPUT TRANSFORMER

TYPE NO.	PRIMARY IMPEDANCE	D.C. Ma	AUDIO WATT	CASE DIMENSIONS H	W	D	MW
S-10X	10000	45	4-6	1$\frac{11}{16}$	2$\frac{11}{16}$	1½	2⅜
S-20X	2000	50	2-3	1$\frac{3}{16}$	2⅛	1¼	1¾
S-40X	14000	5.5	¼	$\frac{13}{16}$	1⅝	⅞	1⅜

COMPONENT OUTLINE

EXAMPLE; OUTPUT TRANSFORMER,
Primary Impedance — 2000
50 Ma D.C., 2-3 Watt,
NO. of Transformer — S-20X

COMPONENT;
LOW VOLTAGE TRANSFORMER, 1 AMP. D.C.
Transformer — F-92A

CASE DIMENSIONS
H — 3½ MW — 2
W — 2$\frac{31}{32}$ MD — 2¼
D — 3
T — 1.0 TD — 2¼

COMPONENT OUTLINE

MANUFACTURER	REF. DESIG.	COMPONENT
TRIAD TRANSFORMER CORPORATION	T	TRANSFORMER

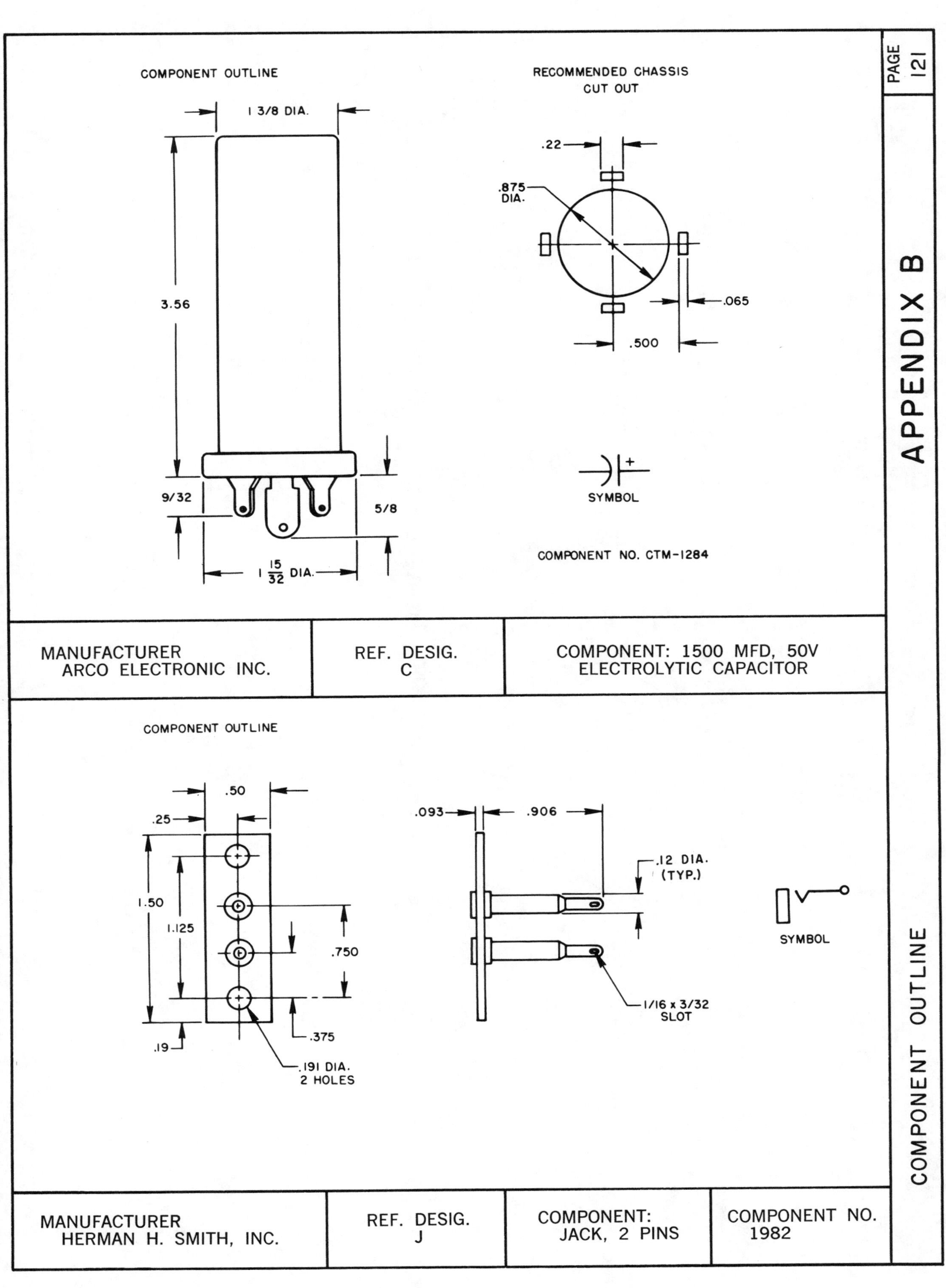
PAGE
121
APPENDIX B
COMPONENT OUTLINE
COMPONENT OUTLINE
1 3/8 DIA.
3.56
9/32
5/8
1 15/32 DIA.
RECOMMENDED CHASSIS
CUT OUT
.22
.875
DIA.
.065
.500
+
SYMBOL
COMPONENT NO. CTM-1284
MANUFACTURER
ARCO ELECTRONIC INC.
REF. DESIG.
C
COMPONENT: 1500 MFD, 50V
ELECTROLYTIC CAPACITOR
COMPONENT OUTLINE
.50
.25
1.50
1.125
.750
.375
.19
.191 DIA.
2 HOLES
.093
.906
.12 DIA.
(TYP.)
1/16 x 3/32
SLOT
SYMBOL
MANUFACTURER
HERMAN H. SMITH, INC.
REF. DESIG.
J
COMPONENT:
JACK, 2 PINS
COMPONENT NO.
1982

COMPONENT:
SWITCHES, TOGGLE
SINGLE POLE

REF. DESIG.
S

FED. SUP CLASS 5930

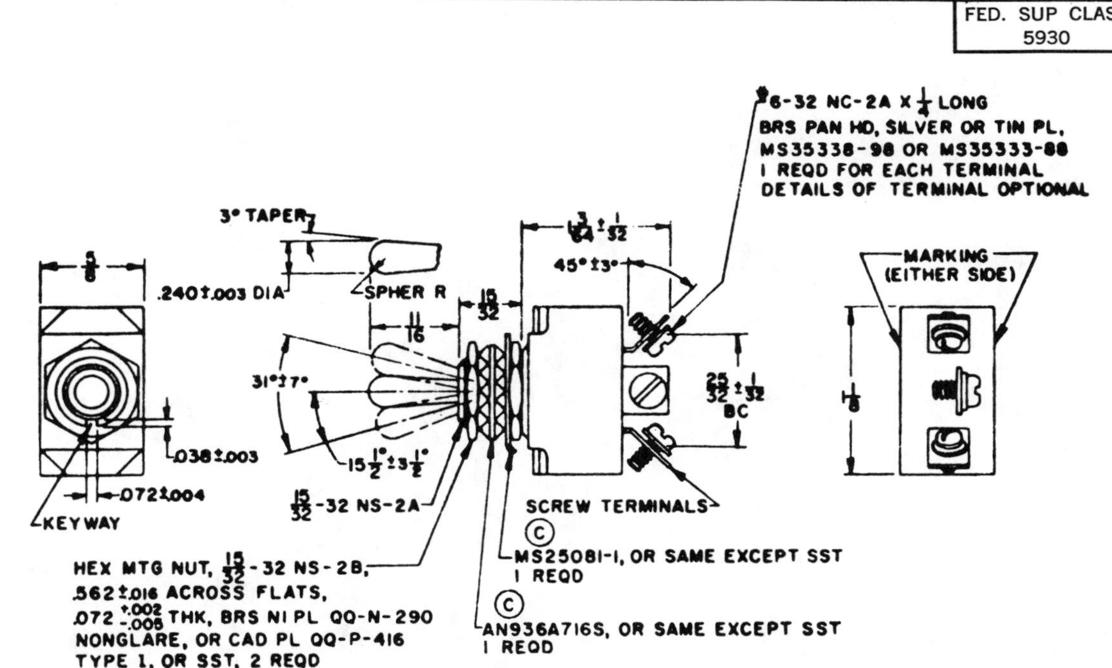

MS PART NO.	FORMER MS PART NO.		FORMER AN PART NO.	FORMER JAN TYPE DESIGNATION	CIRCUIT WITH TOGGLE LEVER IN		
	SCREW-LUG TERMINAL	SOLDER[1]-LUG TERMINAL			Up position	Center position	Down position (keyway side)
MS35058-21	MS35058-5	—	AN3021-1	ST40E	On	Off	On
	—	MS35058-16	—	ST42E			
MS35058-22	—	—	AN3021-2	ST40A	On	None	Off
	—	MS35058-9	—	ST42A			
MS35058-23	MS35058-4	—	AN3021-3	ST40D	On	None	On
	—	MS35058-15	—	ST42D			
MS35058-24	MS35058-1	—	AN3021-10	—	On	Off	None
	—	MS35058-12	—	—			
MS35058-25	MS35058-3	—	AN3021-12	—	On	Mom off	None
	—	MS35058-14	—	—			
MS35038-26	MS35058-6	—	—	ST40F	On	None	Mom on
	—	MS35058-17	—	ST42F			
MS35058-27	MS35058-7	—	AN3021-7	ST40G	Mom on	Off	Mom on
	—	MS35058-18	—	ST42G			
MS35058-28	MS35058-2	—	AN3021-11	—	None	Off	Mom on
	—	MS35058-13	—	—			
MS35058-29	—	—	AN3021-9	ST40B	On	None	Mom off
	—	MS35058-10	—	ST42B			
MS35058-30	—	—	AN3021-8	ST40C	Off	None	Mom on
	—	MS35058-11	, —	ST42C			
MS35058-31	MS35058-8	—	AN3021-6[2]	ST40H	On	Off	Mom on
	—	MS35058-19	—	ST42H			

[1]These MS part numbers are superseded by MS part numbers MS35058-21 to MS35058-31, inclusive, as applicable. By removing the screws and lockwashers from the applicable part, a switch with solder-lug terminals is formed.
[2]With jumper removed.

APPROVED 24 July 1951 REVISED Ⓐ 17 May 1955 Ⓑ 15 April 1958 Ⓒ 20 June 1960

P.A. USAF Other Cust SigC Wep	TITLE SWITCHES, TOGGLE, SINGLE POLE, ONE-HOLE MOUNTING (SEALED TOGGLE BUSHING)	MILITARY STANDARD MS35058
PROCUREMENT SPECIFICATION MIL-S-3950	SUPERSEDES: Air Force — Navy Aeronautical Standard AN3021 dated 12 January 1954	SHEET 1 OF

APPENDIX B

COMPONENT OUTLINE

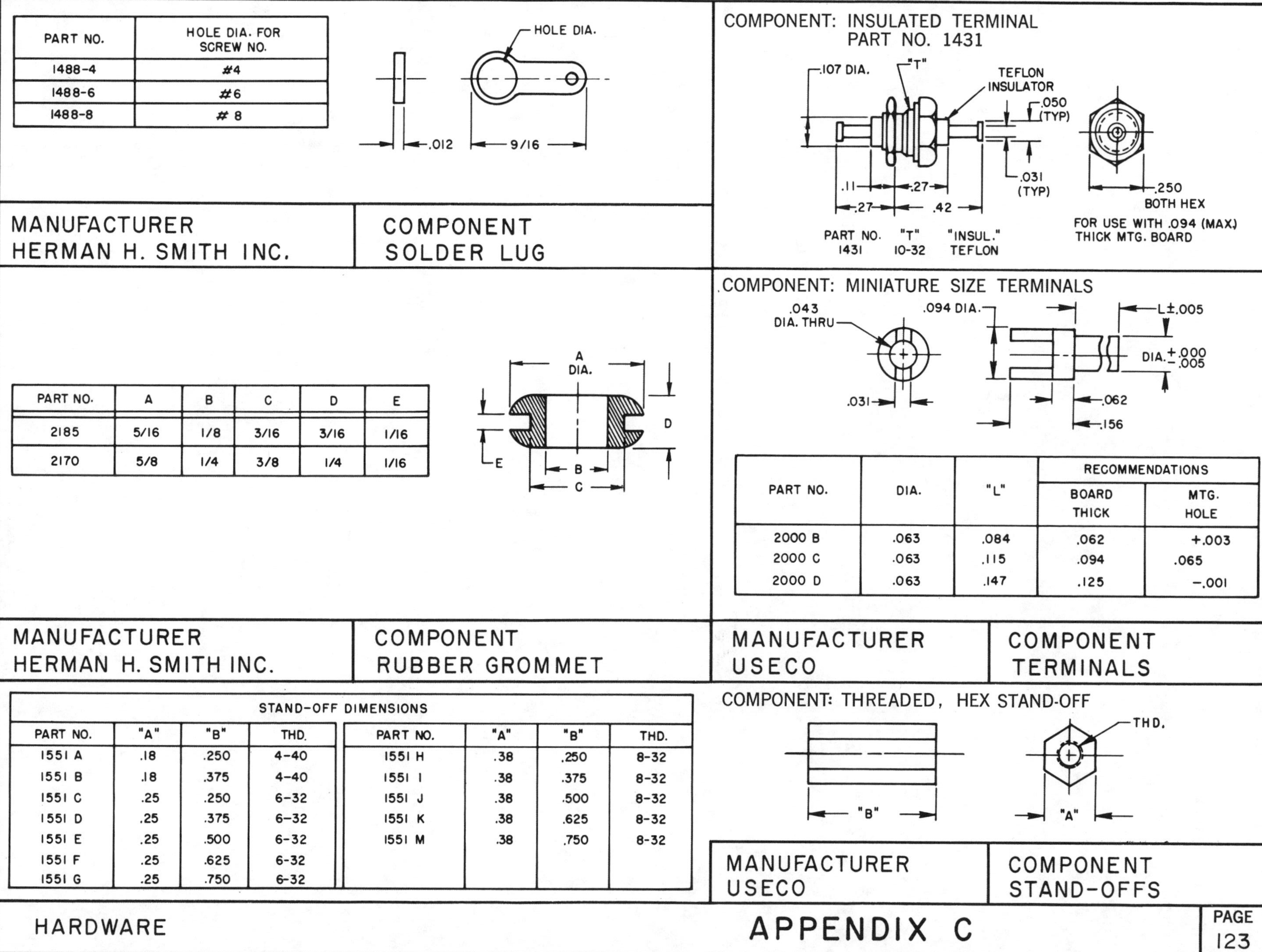

PART NO.	HOLE DIA. FOR SCREW NO.
1488-4	#4
1488-6	#6
1488-8	# 8

MANUFACTURER
HERMAN H. SMITH INC.

COMPONENT
SOLDER LUG

COMPONENT: INSULATED TERMINAL
PART NO. 1431

PART NO.	"T"	"INSUL."
1431	10-32	TEFLON

FOR USE WITH .094 (MAX.) THICK MTG. BOARD

PART NO.	A	B	C	D	E
2185	5/16	1/8	3/16	3/16	1/16
2170	5/8	1/4	3/8	1/4	1/16

MANUFACTURER
HERMAN H. SMITH INC.

COMPONENT
RUBBER GROMMET

COMPONENT: MINIATURE SIZE TERMINALS

PART NO.	DIA.	"L"	RECOMMENDATIONS	
			BOARD THICK	MTG. HOLE
2000 B	.063	.084	.062	+.003
2000 C	.063	.115	.094	.065
2000 D	.063	.147	.125	−.001

MANUFACTURER
USECO

COMPONENT
TERMINALS

STAND-OFF DIMENSIONS

PART NO.	"A"	"B"	THD.
1551 A	.18	.250	4-40
1551 B	.18	.375	4-40
1551 C	.25	.250	6-32
1551 D	.25	.375	6-32
1551 E	.25	.500	6-32
1551 F	.25	.625	6-32
1551 G	.25	.750	6-32
1551 H	.38	.250	8-32
1551 I	.38	.375	8-32
1551 J	.38	.500	8-32
1551 K	.38	.625	8-32
1551 M	.38	.750	8-32

COMPONENT: THREADED, HEX STAND-OFF

MANUFACTURER
USECO

COMPONENT
STAND-OFFS

FED SUP CLASS
5305

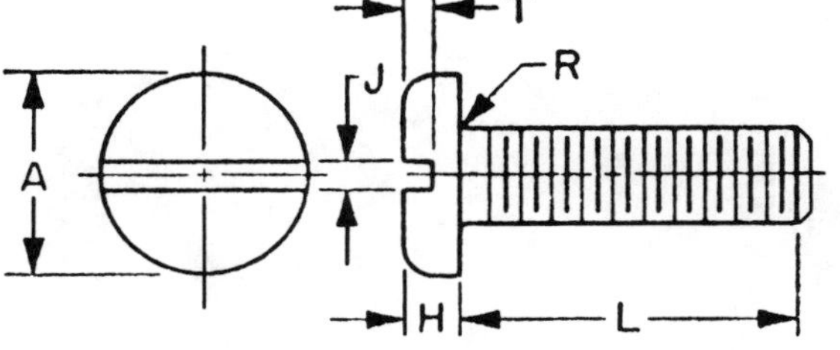

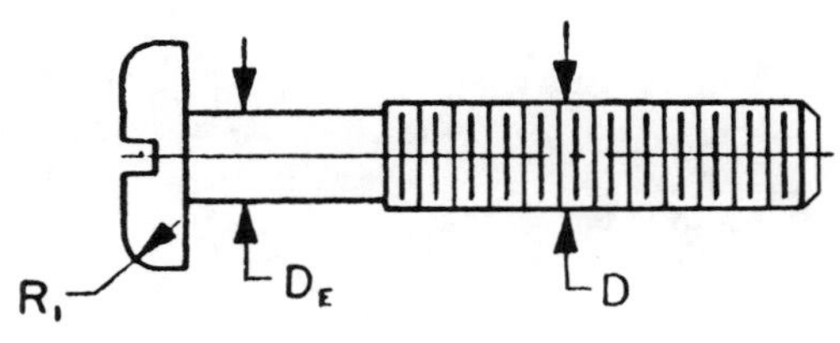

Nominal Size	D		2 (.086)	4 (.112)	6 (.138)	8 (.164)
Threads Per Inch			56NC	40NC	32NC	32NC
Body Diameter	D_E	Max Min	.0860 .0717	.1120 .0925	.1380 .1141	.1640 .1399
Head Diameter	A	Max Min	.167 .155	.219 .205	.270 .256	.322 .306
Head Height	H	Max Min	.053 .045	.068 .058	.082 .072	.096 .085
Slot Width	J	Max Min	.031 .023	.039 .031	.048 .039	.054 .045
Slot Depth	T	Max Min	.033 .023	.041 .030	.050 .038	.058 .043
Radius On Head	R_1	Nom	.035	.042	.046	.052
Radius Under Head	R	Max	.013	.018	.023	.023

Length	L	Tolerance	Dash No.	FIIN	Dash No.	FIIN	Dash No.	FIIN	Dash No.	FIIN
Threads shall extend to within 2 threads of the bearing surface of the head, or closer if practicable.	⅛	+0 −1/32	1		11	043-6472	24	043-6500	39	043-6528
	3/16		2		12	043-6473	25	019-3254	40	043-6529
	¼		3		13	043-6474	26	019-3256	41	043-6530
	5/16		4		14	043-6475	27	043-6501	42	043-6531
	⅜		5		15	043-6476	28	043-6502	43	043-6532
	7/16		6		16	043-6477	29	043-6503	44	043-6533
	½		7		17	043-6478	30	043-6504	45	043-6534
	⅝		8		18	043-6479	31	043-6505	46	043-6535
	¾		9		19	043-6480	32	043-6506	47	043-6536
	⅞		10		20	043-6481	33	043-6507	48	043-6537
	1				21	043-6482	34	043-6508	49	043-6538
	1¼	+0 −1/16			22	043-6483	35	043-6509	50	043-6539
	1½				23	043-6486	36	043-6510	51	043-6540
	1¾						37	043-6511	52	043-6541
	2						38	043-6512	53	043-6542
Minimum complete thread length of 1¾.	2¼	+0 −3/32							54	043-6543
	2½								55	043-6544
	2¾								56	043-6545
	3								57	043-6546

Materials: Carbon Steel, Specification QQ-W-409 or QQ-S-633, except Bessemer Compositions; 55,000 PSI minimum ultimate tensile strength.

Finish: Plain (untreated).

Threads: The threads shall be in accordance with Screw-thread Standards for Federal Services Handbook H-28.

Notes:
(1) Referenced documents of issue in effect on date of invitation for bids shall apply.
(2) In case of conflict with any referenced document, this standard will govern.
(3) The MS part number consists of the MS sheet number, plus the dash number. Example: MS 35221-1.
(4) All dimensions in inches.

APPROVED FEB 9 1954 REVISED

CUSTODIANS	OTHER INT.	MILITARY STANDARD	MS35221
A — ORD N — SHIPS AF — USAF	A — CEMQSIGT N — AMDORSYMC AF —	SCREW, MACHINE, PAN HEAD, SLOTTED, CARBON STEEL, PLAIN FINISH, NC-2A AND UNC-2A	
PROCUREMENT SPECIFICATION FF — S — 92		SUPERSEDES:	SHEET 1 OF 2

APPENDIX C

FED SUP CLASS 5310

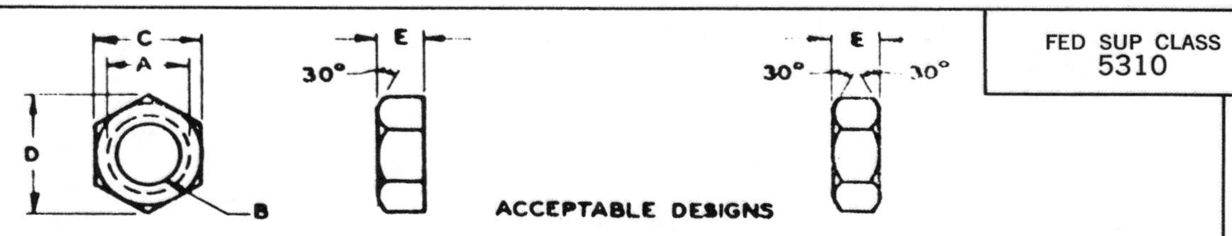

			No. 2	No. 4	No. 6	No. 8
A	Nominal Size or Basic Major Dia. of Thread		No. 2 .086	No. 4 .112	No. 6 .138	No. 8 .164
B	Threads per Inch		56	40	32	32
C	Width Across Flats	Nom Max Min	3/16 .1875 .180	1/4 .2500 .241	5/16 .3125 .302	11/32 .3438 .332
D	Width Across Corners	Max Min	.217 .205	.289 .275	.361 .344	.397 .378
E	Thickness	Nom Max Min	1/16 .066 .057	3/32 .098 .087	7/64 .114 .102	1/8 .130 .117

Material and Protective Coating	Dash No.	FIIN	Dash No.	FIIN	Dash No.	FIIN	Dash No.	FIIN
Steel, Carbon								
Uncoated	21	019-1716	41	011-4776	61	275-1706	81	275-6800
Cadmium or Zinc Optional	22		42	013-4524	62	013-4530	82	012-0622
Phosphate	23		43	275-9301	63		83	275-9310
Steel, Corrosion Resisting								
Passivated	24	271-4640	44	271-4642	64	271-4644	84	271-4645
Brass								
Uncoated	25		45		65		85	
Tin Plated	26		46		66		86	

A	Nominal Size or Basic Major Dia. of Thread		No. 10 .190			
B	Threads per Inch		24			
C	Width Across Flats	Nom Max Min	3/8 .3750 .362			
D	Width Across Corners	Max Min	.433 .413			
E	Thickness	Nom Max Min	1/8 .130 .117			

Material and Protective Coating	Dash No.	FIIN	Dash No.	FIIN	Dash No.	FIIN	Dash No.	FIIN
Steel, Carbon								
Uncoated	101	350-3384						
Cadmium or Zinc Optional	102	012-0361						
Phosphate	103	281-5341						
Steel, Corrosion Resisting								
Passivated	104	275-5095						
Brass								
Uncoated	105							
Tin Plated	106							

Material: Steel, Carbon, (Commercial Grade) except Bessemer Steels.
Steel, Corrosion Resisting, Federal Standard No. 66, Steel Numbers: #303, 304, 305, 410, 416, 430.
Brass, Naval, Specification MIL-B-994, Composition A or C.

Protective Coating: Cadmium Plate, Specification QQ-P-416, Type II, Class C.
Zinc Plate, Specification QQ-Z-325, Type II, Class 3.
Phosphate, Specification MIL-C-16232, Type II.
Tin Plate, Specification MIL-T-10727, Type I or II, .0001 thick.

Thread: The threads shall be in accordance with Screw-Thread Standards for Federal Services, Handbook H-28.

Notes:
(1) Referenced documents shall be of the issue in effect on date of invitations for bid.
(2) This document has been promulgated by the Department of Defense as the Military Standard to limit the selection of the item, product or design covered herein in engineering, design and procurement. This standard shall become effective not later than 90 days after the latest date of approval shown.
(3) This standard takes precedence over documents referenced herein.
(4) Nuts shall be free from burrs, scale and all other defects that would affect their serviceability.
(5) The MS part number consists of the MS sheet number, plus the dash number. Example: MS35649-22.
(6) All dimensions in inches.

APPROVED 22 DEC 55 REVISED

CUSTODIANS	OTHER INT.	MILITARY STANDARD	
A — ORD N — SHIPS AF — USAF	A — CESIGT N — MCORY AF —	NUT, PLAIN, HEXAGON, MACHINE SCREW, NC-2B	MS35649
PROCUREMENT SPECIFICATION FF-N-836		SUPERSEDES:	SHEET 1 OF 1

FED SUP CLASS 5310

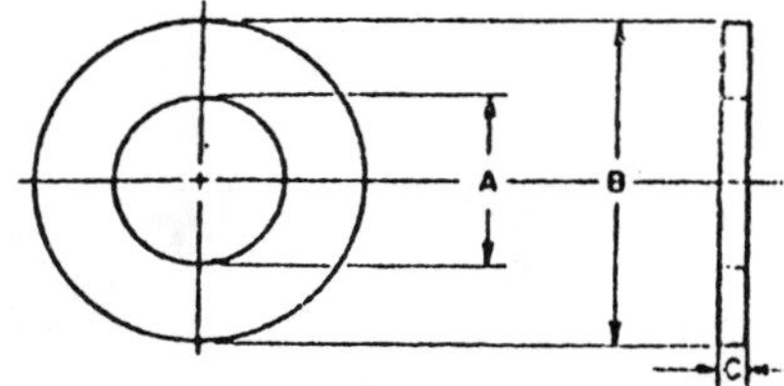

TOLERANCES

±.010 ON OUTSIDE DIAMETER.
±.005 ON INSIDE DIAMETER UP TO AND INCLUDING #10.
±.010 ON ALL OTHER INSIDE DIAMETERS.
INSIDE AND OUTSIDE DIAMETERS SHALL BE CONCENTRIC WITHIN THE TOLERANCE OF THE INSIDE DIAMETER.

NOM. SIZE	"A" I.D.	"B" O.D.	"C" THICKNESS		CRES		Ni-Cu ALLOY		COPPER		BRASS		ALUMINUM ALLOY	
			Max.	Min.	Dash No.	FIIN	Dash No.	FIIN	Dash No.	FIIN	Dash No.	FIIN	Dash No.	FIIN
0	.078	.187	.025	.016	301		401		501		601		701	
2	.093	.250	.025	.016	302		402		502		602		702	
4	.125	.250	.028	.017	303		403		503		603		703	
4	.125	.312	.040	.025	304		404		504		604		704	
6	.156	.312	.048	.027	305		405		505		605		705	
6	.156	.375	.065	.036	306		406		506		606		706	
8	.187	.375	.065	.036	307		407		507		607		707	
10	.218	.437	.065	.036	308		408		508		608		708	
10	.250	.562	.080	.051	309		409		509		609		709	
¼	.281	.625	.080	.051	310		410		510		610		710	
¼	.312	.750	.080	.051	311		411		511		611		711	
5/16	.343	.687	.080	.051	312		412		512		612		712	
5/16	.375	.875	.104	.064	313		413		513		613		713	
⅜	.406	.812	.080	.051	314		414		514		614		714	
⅜	.437	1.000	.104	.064	315		415		515		615		715	
7/16	.468	.921	.080	.051	316		416		516		616		716	
7/16	.500	1.250	.104	.064	317		417		517		617		717	
½	.531	1.062	.121	.074	318		418		518		618		718	
½	.562	1.375	.132	.086	319		419		519		619		719	
⅝	.656	1.312	.121	.074	320		420		520		620		720	
⅝	.687	1.750	.160	.108	321		421		521		621		721	
¾	.812	1.500	.160	.108	322		422		522		622		722	
¾	.812	2.000	.177	.122	323		423		523		623		723	
⅞	.937	1.750	.160	.108	324		424		524		624		724	
⅞	.937	2.250	.192	.136	325		425		525		625		725	
1	1.062	2.000	.160	.108	326		426		526		626		726	
1	1.062	2.500	.192	.136	327		427		427		627		727	
1⅛	1.250	2.750	.192	.136	328		428		528		628		728	
1¼	1.375	3.000	.192	.136	329		429		529		629		729	
1⅜	1.500	3.250	.213	.153	330		430		530		630		730	
1½	1.625	3.500	.213	.153	331		431		531		631		731	
1⅝	1.750	3.750	.213	.153	332		432		532		632		732	
1¾	1.875	4.000	.213	.153	333		433		533		633		733	
1⅞	2.000	4.250	.213	.153	334		434		534		634		734	
2	2.125	4.500	.213	.153	335		435		535		635		735	
2¼	2.375	4.750	.248	.193	336		436		536		636		736	
2½	2.625	5.000	.280	.210	337		437		537		637		737	
2¾	2.875	5.250	.310	.228	338		438		538		638		738	
3	3.125	5.500	.327	.249	339		439		539		639		739	

All dimensions are in inches.

MATERIALS: Steel, Corrosion Resisting; QQ-S-765, 60,000 psi minimum tensile strength, 1% minimum elongation in 2 inches.
Nickel — Copper Alloy (Monel); QQ-N-281, Class A
Copper; QQ-C-576
Brass, Half Hard, QQ-B-611, Composition C
Aluminum Alloy, Half Hard; QQ-A-359

PROTECTIVE COATINGS:
Corrosion Resisting Steel Washers shall be passivated.
Aluminum Alloy Washers shall be anodized in accordance with MIL-A-8625 or given a chemical film in accordance with MIL-C-5541.

IDENTIFICATION NUMBER — (MS Number) — (Washer Dash No.)
Example: MS 15795 — 318 would be the identification number of a ½ nominal size corrosion resisting steel washer with a 1.062 outside diameter.

NOTES: 1. In case of conflict with any referenced document this standard will govern.
2. Referenced documents shall be the issue in effect on the date at invitation for bids.

APPROVED 2 June 1954 REVISED

CUSTODIANS	OTHER INT.	MILITARY STANDARD	MS15795
Ord NOrd USAF	A — CESIGT N — ASHYMC AF —	WASHERS, FLAT, METAL, ROUND, GENERAL PURPOSE	
PROCUREMENT SPECIFICATION NONE		SUPERSEDES:	SHEET 2 OF 2

NOMINAL SIZE	INSIDE DIAMETER — A —		WIDTH — W —	THICKNESS — C — $\frac{T+t}{2}$		OUTSIDE DIAMETER — B —	STEEL								CORROSION RESISTING		PHOSPHOR BRONZE	
							PLAIN (UNCOATED)		CADMIUM OR ZINC OPTIONAL		CADMIUM		PHOSPHATE		PASSIVATED		CADMIUM	
	MAX.	MIN.	MIN.	MAX.	MIN.	MAX.	Dash No.	FIIN	Dash No.	FIIN	Dash No.	FIIN	Dash No.	FIIN	Dash No.	FIIN	Dash No.	FIIN
#2 .086	.097	.088	.030	.021	.015	.165	1	019-2298	20		39		58	019-2303	77	058-2950	96	
#4 .112	.124	.115	.035	.026	.020	.202	2	011-8871	21		40		59	013-1195	78	058-2949	97	
#6 .138	.151	.141	.040	.031	.025	.239	3	011-8872	22		41		60	013-1196	79	043-1754	98	
#8 .164	.178	.168	.047	.037	.031	.280	4	011-8869	23		42		61	013-1197	80	042-9067	99	
#10 .190	.205	.194	.055	.046	.040	.323	5	011-8873	24		43		62	274-8708	81	058-2951	100	
¼	.267	.255	.107	.057	.047	.489	6	011-3114	25		44		63	274-8714	82	043-5862	101	
5/16	.333	.319	.117	.066	.056	.575	7	011-2723	26		45		64	013-1201	83		102	
⅜	.398	.382	.136	.080	.070	.678	8	011-0730	27		46		65	274-8719	84		103	
7/16	.464	.446	.154	.095	.085	.780	9	011-0405	28		47		66	012-1739	85		104	
½	.529	.509	.170	.109	.099	.877	10	010-6500	29		48		67	013-1203	86		105	261-7125
9/16	.595	.573	.186	.123	.113	.975	11	011-2724	30		49		68		87		106	
⅝	.660	.636	.201	.136	.126	1.082	12	010-3334	31		50		69	012-3167	88		107	189-6811
¾	.791	.763	.233	.163	.153	1.277	13	010-3335	32		51		70	013-1226	89		108	
⅞	.922	.890	.264	.199	.179	1.470	14	010-3336	33		52		71	013-1227	90		109	
1	1.053	1.017	.289	.222	.202	1.656	15	011-7661	34		53		72		91		110	
1⅛	1.184	1.144	.314	.244	.224	1.837	16	187-3202	35		54		73	013-1229	92		111	
1¼	1.315	1.271	.336	.264	.244	2.012	17	011-7613	36		55		74		93		112	
1⅜	1.446	1.398	.356	.284	.264	2.183	18	011-8028	37		56		75		94		113	
1½	1.577	1.525	.375	.302	.282	2.352	19	011-8029	38		57		76		95		114	

Material: Steel, Carbon, FS1060 to FS1080, Rockwell "C" 45-53, Specification QQ-S-633.
Corrosion Resisting Steel, Federal Standard No. 66, Steel Numbers 302 Rockwell "C" 35-43 or 420 Rockwell "C" 43-53.
Phosphor Bronze, Specification QQ-B-746, Composition A, Hard.

Protective Coating:
Cadmium Plate, Specification QQ-P-416, Type II, Class C.
Zinc Plate, Specification QQ-Z-325, Type II, Class 3.
Phosphate, Specification MIL-C-16232, Type II.

Dimensions: All dimensions are in inches unless otherwise specified.
Part Numbers: The MS part number consists of the MS number, plus the dash number. Example: MS35337-1.
Notes:
(1) Referenced documents shall be of the issue in effect on invitations for bid.
(2) This standard takes precedence over documents referenced herein.
(3) This document has been promulgated by the Department of Defense as the Military Standard to limit the selection of the item, product or design covered herein in engineering, design and procurement. This standard shall become effective not later than 90 days after the latest date of approval shown.

FED SUP CLASS 5310

APPROVED MAR 4 1954 REVISED 28 APRIL 56

CUSTODIANS
A — Ord
N — Ships
AF — AF

OTHER INT.
A — CESIGT
N — MCORASY
AF —

PROCUREMENT SPECIFICATION FF-W-84

MILITARY STANDARD

WASHER, LOCK, SPLIT, HELICAL, LIGHT SERIES

SUPERSEDES:

MS35337

SHEET 1 OF 1

SCREW CLEARANCE AND HOLE CHART

SCREW NO.	SCREW BODY DIA. (d)	SINGLE CLEARANCE HOLE DIA.	MINIMUM SHEET THICKNESSES FOR 82° AND 100° CSK. HOLES		CSK. DIA. ±.005
			82°	100°	
2	.086	.089 (#43)	.063		.151
4	.112	.120 (#31)	.083	.064	.225
6	.138	.144 (#27)	.095	.072	.289
8	.164	.172 (11/64)	.109	.081	.337
10	.190	.194 (#11)	.125	.091	.390
1/4	.250	.257 (F)	.168	.125	.512
5/16	.313	.316 (O)	.209	.156	.640
3/8	.375	.386 (W)	.253	.188	.767
7/16	.438	.453 (29/64)	.241	.204	.895
1/2	.500	.516 (33/64)	.241	.231	1.002
9/16	.563	.578 (37/64)	.271	.251	1.150
5/8	.625	.641 (41/64)	.312	.286	1.277

FOR MULTIPLE HOLE PATTERN USE THE FOLLOWING FORMULAS:*
FOR CLEARANCE HOLE ON TAPPED ONLY

2 HOLE PATTERN $D = d + 2t$
3 HOLE PATTERN $D = d + 4t$
4 HOLE PATTERN $D = d + 2.82t$
6 HOLE PATTERN $D = d + 5.62t$
OR MORE THAN 6 HOLES.

WHERE D = CLEARANCE HOLE DIA.
d = SCREW BODY DIA.
t = TOLERANCE (℄ to ℄ HOLE)

FOR CLEARANCE HOLE ON CLEARANCE HOLE DIVIDE THE LAST PART BY 2

$\frac{2t}{2}$, $\frac{4t}{2}$ etc.

CSK. ANGLE
CSK. DIA.
SHEET THICKNESS
HOLE DIA.

EXAMPLE:

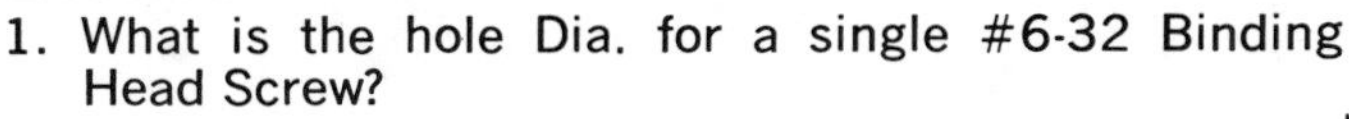
1. What is the hole Dia. for a single #6-32 Binding Head Screw?

ANSWER: $.144 {}^{+.005}_{-.001}$ DIA.

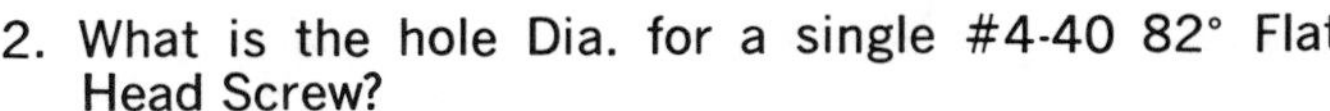
2. What is the hole Dia. for a single #4-40 82° Flat Head Screw?

ANSWER: $.120 {}^{+.004}_{-.001}$ DIA. HOLE
CSINK .225 DIA. x 82°

3. What hole Dia. should be drilled for a #4-40 Binding Head Screw, 4 Places, (Tolerance between holes ℄ to ℄ is ±.005) Tapped Hole Mounted.

ANSWER; $D = d + 2.82t$
$D = .112 + (2.82 \times .005)$
$D = .112 + .0141 = .1261$

Therefore Hole Dia. should read to the next larger Drill Size (See Appendix E)

$.128 {}^{+.005}_{-.001}$ DIA.

4 HOLES

STANDARD DRILLED HOLE TOLERANCES	
HOLE DIA.	TOLERANCE
.0135 THRU .125	+.004 −.001
.126 THRU .250	+.005 −.001
.251 THRU .500	+.006 −.001
.501 THRU .750	+.008 −.001
.751 THRU 1.000	+.010 −.001
1.001 THRU 2.000	+.012 −.001

*For exercises see Lesson NO. 1 "MECHANICAL DRAFTING REVIEW."

FRACTIONS AND DECIMAL EQUIVALENTS

Fraction	Decimal	Fraction	Decimal
1/64	.0156	33/64	.5156
1/32	.0312	17/32	.5312
3/64	.0468	35/64	.5468
1/16	.0625	9/16	.5625
5/64	.0781	37/64	.5781
3/32	.0937	19/32	.5937
7/64	.1093	39/64	.6093
1/8	.125	5/8	.625
9/64	.1406	41/64	.6406
5/32	.1562	21/32	.6562
11/64	.1718	43/64	.6718
3/16	.1875	11/16	.6875
13/64	.2031	45/64	.7031
7/32	.2187	23/32	.7187
15/64	.2343	47/64	.7343
1/4	.25	3/4	.75
17/64	.2656	49/64	.7656
9/32	.2812	25/32	.7812
19/64	.2968	51/64	.7968
5/16	.3125	13/16	.8125
21/64	.3281	53/64	.8281
11/32	.3437	27/32	.8437
23/64	.3593	55/64	.8593
3/8	.375	7/8	.875
25/64	.3906	57/64	.8906
13/32	.4062	29/32	.9062
27/64	.4218	59/64	.9218
7/16	.4375	15/16	.9375
29/64	.4531	61/64	.9531
15/32	.4687	31/32	.9687
31/64	.4843	63/64	.9843
1/2	.5	1	1.

OHM'S LAW

VOLTS(E)	OHMS(R)	AMPERES(I)	WATTS(W)
$E = IR$	$R = E/I$	$I = E/R$	$W = EI$
$E = \sqrt{WR}$	$R = E^2/W$	$I = W/E$	$W = I^2R$
$E = W/I$	$R = W/I^2$	$I = \sqrt{W/R}$	$W = E^2R$

DRILL SIZES — DECIMAL EQUIVALENTS

Drill Size	Decimal Equivalent
80	.0135
79	.0145
1/64	.0156
78	.0160
77	.0180
76	.0200
75	.0210
74	.0225
73	.0240
72	.0250
71	.0260
70	.0280
69	.0292
68	.0310
1/32	.0313
67	.0320
66	.0330
65	.0350
64	.0360
63	.0370
62	.0380
61	.0390
60	.0400
59	.0410
58	.0420
57	.0430
56	.0465
3/64	.0469
55	.0520
54	.0550
53	.0595
1/16	.0625
52	.0635
51	.0670
50	.0700
49	.0730
48	.0760
5/64	.0781
47	.0785
46	.0810
45	.0820
44	.0860
43	.0890
42	.0935
3/32	.0937
41	.0960
40	.0980
39	.0995
38	.1015
37	.1040
36	.1065
7/64	.1093
35	.1100
34	.1110
33	.1130
32	.1160
31	.1200

Drill Size	Decimal Equivalent
1/8	.1250
30	.1285
29	.1360
28	.1405
9/64	.1406
27	.1440
26	.1470
25	.1495
24	.1520
23	.1540
5/32	.1562
22	.1570
21	.1590
20	.1610
19	.1660
18	.1695
11/64	.1719
17	.1730
16	.1770
15	.1800
14	.1820
13	.1850
3/16	.1875
12	.1890
11	.1910
10	.1935
9	.1960
8	.1990
7	.2010
13/64	.2031
6	.2040
5	.2055
4	.2090
3	.2130
7/32	.2187
2	.2210
1	.2280
A	.2340
15/64	.2344
B	.2380
C	.2420
D	.2460
1/4E	.2500
F	.2570
G	.2610
17/64	.2656
H	.2660
I	.2720
J	.2770
K	.2811
9/32	.2812
L	.2900
M	.2950
19/64	.2968
N	.3020
5/16	.3125

Drill Size	Decimal Equivalent
O	.3160
P	.3230
21/64	.3281
Q	.3320
R	.3390
11/32	.3437
S	.3480
T	.3580
23/64	.3594
U	.3680
3/8	.3750
V	.3770
W	.3860
25/64	.3906
X	.3970
Y	.4040
13/32	.4062
Z	.4130
27/64	.4219
7/16	.4375
29/64	.4531
15/32	.4687
31/64	.4843
1/2	.5000
33/64	.5156
17/32	.5313
35/64	.5469
9/16	.5625
37/64	.5781
19/32	.5937
39/64	.6094
5/8	.6250
41/64	.6406
21/32	.6562
43/64	.6719
11/16	.6875
45/64	.7031
23/32	.7187
47/64	.7344
3/4	.7500
49/64	.7656
25/32	.7812
51/64	.7969
13/16	.8125
53/64	.8281
27/32	.8437
55/64	.8594
7/8	.8750
57/64	.8906
29/32	.9062
59/64	.9219
15/16	.9375
61/64	.9531
31/32	.9687
63/64	.9844
1	1.0000

COLOR CODE

COLOR	ABBREV.*	NO.	MULTIPLIER	TOL. ±%
BLACK	(BLK) BK	0		
BROWN	(BRN) BR	1	10	
RED	(RED) R	2	100	
ORANGE	(ORN) O	3	1000	
YELLOW	(YEL) Y	4	10^4	CATHODE
GREEN	(GRN) GN	5	10^5	
BLUE	(BLU) BL	6	10^6	ANODE
VIOLET OR PURPLE	(VIO) V (PR) P	7	10^7	
GRAY	(GY) GY	8	10^8	
WHITE	(WHT) W	9	10^9	
GOLD			.1	5%
SILVER			.01	10%
NO COLOR				20%

*ABBREVIATIONS shown in brackets are MILITARY STANDARDS.

*ABBREVIATIONS not shown in brackets were made up specifically for this exercise book.

EXAMPLE: Resistor value by COLOR

1st digit	2nd digit	3rd digit (multiplier)	4th digit (tolerance)
RED 2	YELLOW 4	ORANGE 3	GOLD 5%

Answer; 24000 or 24K, ±5%

(K = 1000, M = 1000000)

APPENDIX E

LOGIC SYMBOLS

In the following list of logic symbols, explanations accompany the graphic representations.

LOGIC SYMBOL	EXPLANATION

AND

INPUT SIDE OUTPUT SIDE

A B F

A B C F

The symbol shown represents the AND function.

The AND output is high (H) if and only if all the inputs are high. A **small circle** at the output indicates **an opposite** output from that of the statement and table above. See lower table at right. The small circle shall never be drawn by itself on a diagram.

TRUTH TABLE

H=High L=Low

Input A	Input B	Output F
L	L	L
L	H	L
H	L	L
H	H	H

AND function with Small Circle

Input A	Input B	Input C	Output F
L	L	L	H
L	L	H	H
L	H	L	H
L	H	H	H
H	L	L	H
H	L	H	H
H	H	L	H
H	H	H	L

INCLUSIVE OR

INPUT SIDE OUTPUT SIDE

A B F

A B C F

The symbol shown represents the INCLUSIVE OR function.

The OR output is high (H) if and only if any one or more of the inputs are high.

Inputs can be drawn on any side of the symbol except the output side.

Input A	Input B	Output F
L	L	L
L	H	H
H	L	H
H	H	H

Input A	Input B	Input C	Output F
L	L	L	L
L	L	H	H
L	H	L	H
L	H	H	H
H	L	L	H
H	L	H	H
H	H	L	H
H	H	H	H

EXCLUSIVE OR

A B F

The symbol shown represents the EXCLUSIVE OR function.

The EXCLUSIVE OR output is high (H) if and only if any one input is high and all other inputs are low.

Input A	Input B	Output F
L	L	L
L	H	H
H	L	H
H	H	L

FLIP-FLOP

S T C

FF

1 0

Aspect ratio = 1.75 : 1

The flip-flop is a device which stores a single bit of information.

It has three possible inputs, set (S), clear (reset) (C), and toggle (trigger) (T), and two possible outputs, 1 and 0.

When not used, the trigger input may be omitted.

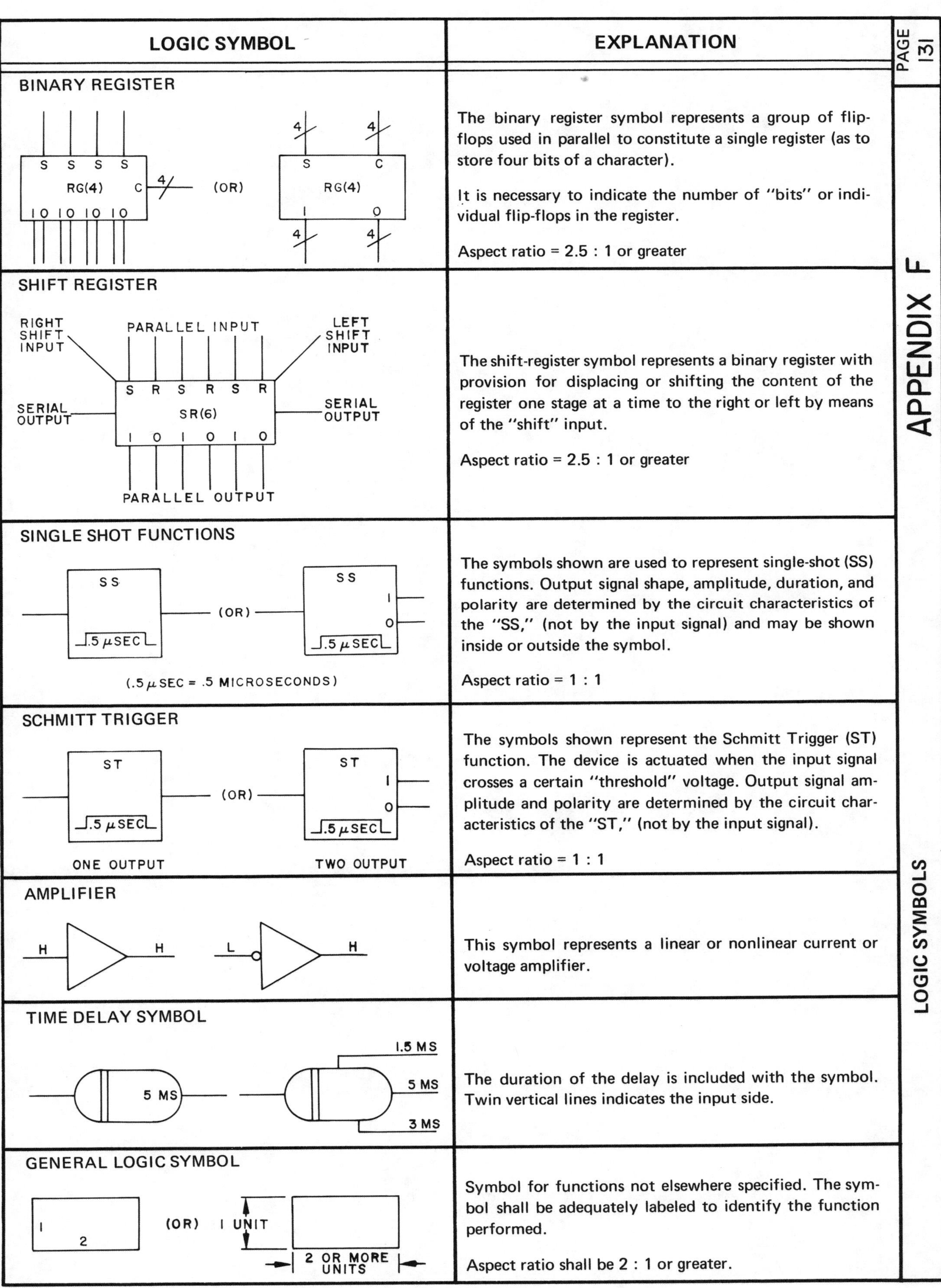

LOGIC SYMBOL	EXPLANATION
BINARY REGISTER	The binary register symbol represents a group of flip-flops used in parallel to constitute a single register (as to store four bits of a character). It is necessary to indicate the number of "bits" or individual flip-flops in the register. Aspect ratio = 2.5 : 1 or greater
SHIFT REGISTER	The shift-register symbol represents a binary register with provision for displacing or shifting the content of the register one stage at a time to the right or left by means of the "shift" input. Aspect ratio = 2.5 : 1 or greater
SINGLE SHOT FUNCTIONS	The symbols shown are used to represent single-shot (SS) functions. Output signal shape, amplitude, duration, and polarity are determined by the circuit characteristics of the "SS," (not by the input signal) and may be shown inside or outside the symbol. Aspect ratio = 1 : 1
SCHMITT TRIGGER	The symbols shown represent the Schmitt Trigger (ST) function. The device is actuated when the input signal crosses a certain "threshold" voltage. Output signal amplitude and polarity are determined by the circuit characteristics of the "ST," (not by the input signal). Aspect ratio = 1 : 1
AMPLIFIER	This symbol represents a linear or nonlinear current or voltage amplifier.
TIME DELAY SYMBOL	The duration of the delay is included with the symbol. Twin vertical lines indicates the input side.
GENERAL LOGIC SYMBOL	Symbol for functions not elsewhere specified. The symbol shall be adequately labeled to identify the function performed. Aspect ratio shall be 2 : 1 or greater.

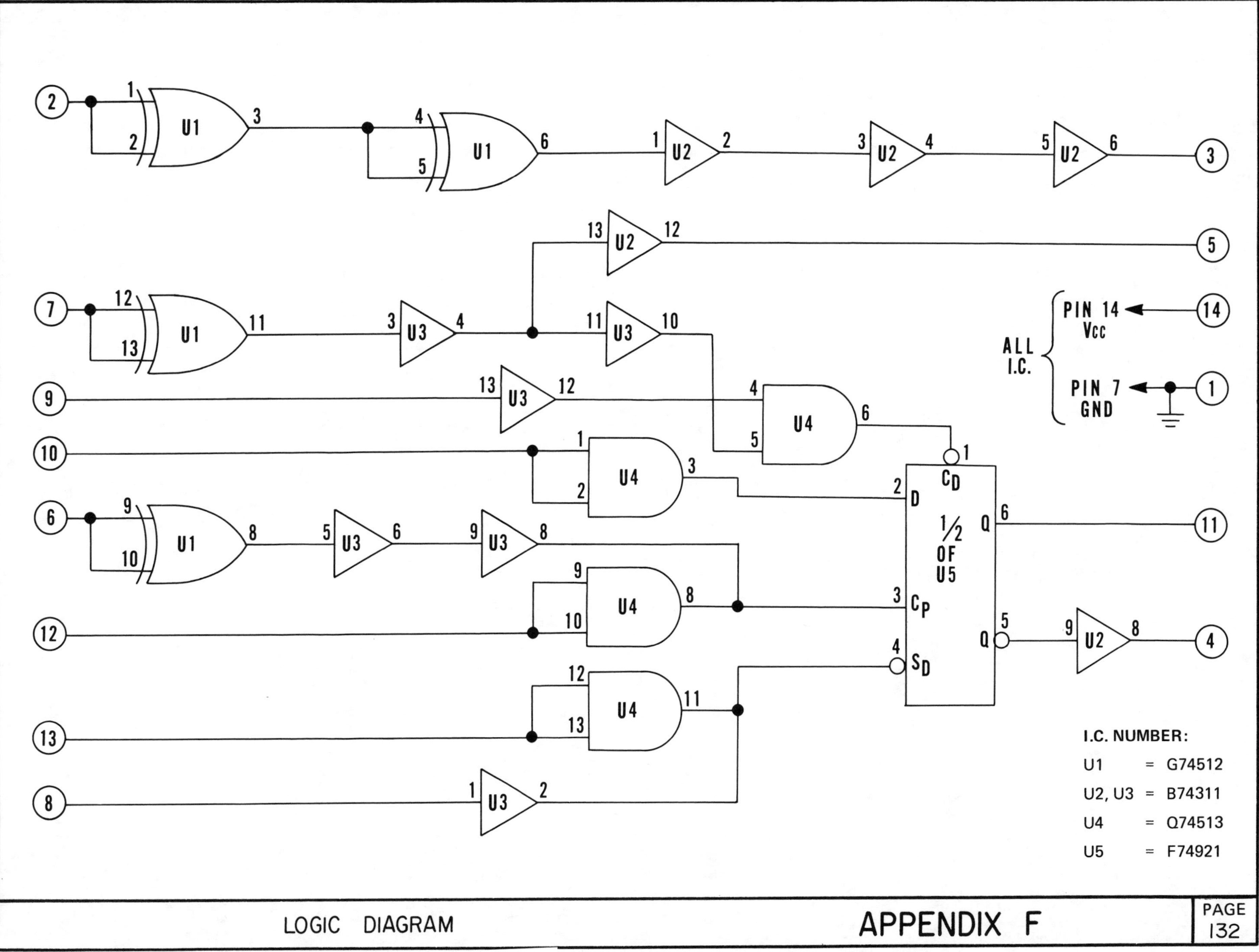

LOGIC DIAGRAM

APPENDIX F

14-PIN DUAL-IN-LINE I.C.

The illustration (Fig. 1) below shows a simple method of mounting a dual-in-line to a printed circuit board.

Pad configurations may vary in shape and O.D. size. The O.D. pad size depends on the pattern of conductor lines crossing on top and bottom of the board.

A common practice is to use the following:

FOR	USE
FULL SCALE layout	.062 O.D. Pad
2 x SIZE layout	.125 O.D. Pad
4 x SIZE layout	.250 O.D. Pad

Due to a wide variation in tolerance of both the diameter and the ℄ to ℄ distance of the dual-in-line leads, a .031 dia. hole, as shown, is recommended for ease in mounting the dual-in-line package.

Plated-thru holes are required when the conductor pattern is both on the top and bottom of the P.C.B.

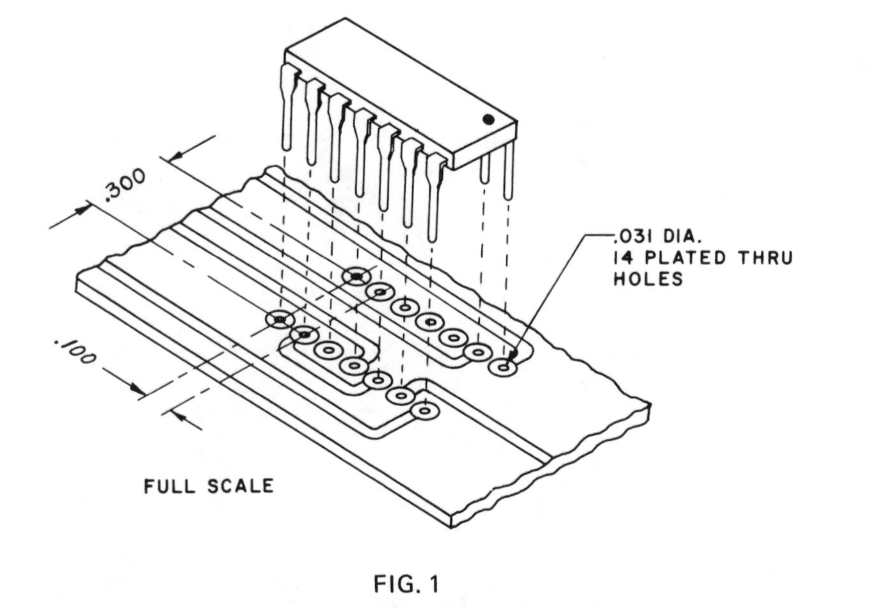

FIG. 1

FIG. 2 14-PIN DUAL-IN-LINE PACKAGE

FIG. 3 I.C. CHIPS

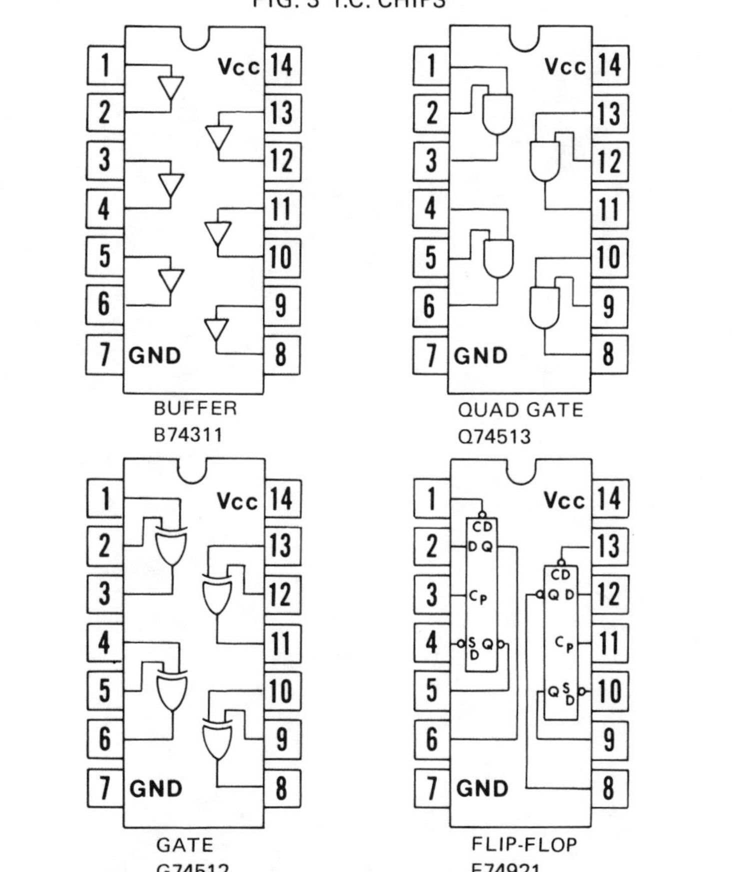

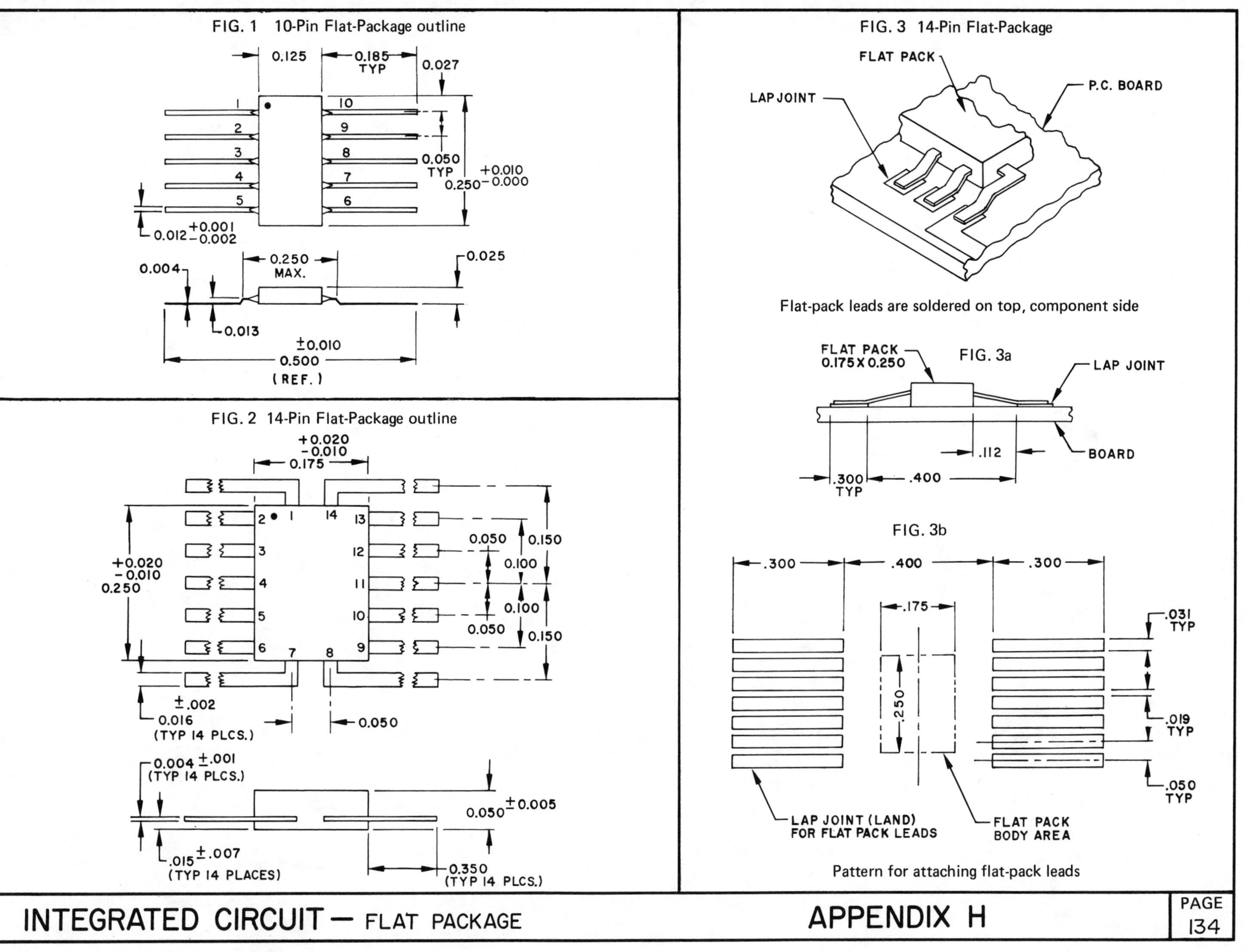

FIG. 1 10-Pin Flat-Package outline

FIG. 2 14-Pin Flat-Package outline

FIG. 3 14-Pin Flat-Package

Flat-pack leads are soldered on top, component side

FIG. 3a

FIG. 3b

Pattern for attaching flat-pack leads

INTEGRATED CIRCUIT — FLAT PACKAGE

APPENDIX H

PAGE 134

A	B	C	D	E
AMPLIFIER OR BUFFER	AND GATE	INPUT/OUTPUT TERMINAL	JUNCTION POINT	EXCLUSIVE OR GATE
F	**G**	**H**	**I**	**J**
FUSE	GROUND EARTH CONNECTION	GROUND CHASSIS CONNECTION	INDUCTOR	INCLUSIVE OR GATE
K	**L**	**M**	**N**	**O**
ADJUSTABLE RESISTOR	GENERAL ANTENNA	DIPOLE ANTENNA	VARIABLE CAPACITOR	SPEAKER
P	**Q**	**R**	**S**	**T**
CAPACITOR	NPN TRANSISTOR	RESISTOR	DIODE	BREAKDOWN DIODE (UNIDIRECTIONAL)
U	**V**	**W**	**X**	**Y**
CAPACITOR	RESISTOR	NPN TRANSISTOR	DIODE	CAPACITOR

ELECTRONIC SYMBOLS MENU #ELSYM1

NOTE 1

Menu entries U, V, W, X, and Y to be made by student as per exercise of CAD-6 (page 89).

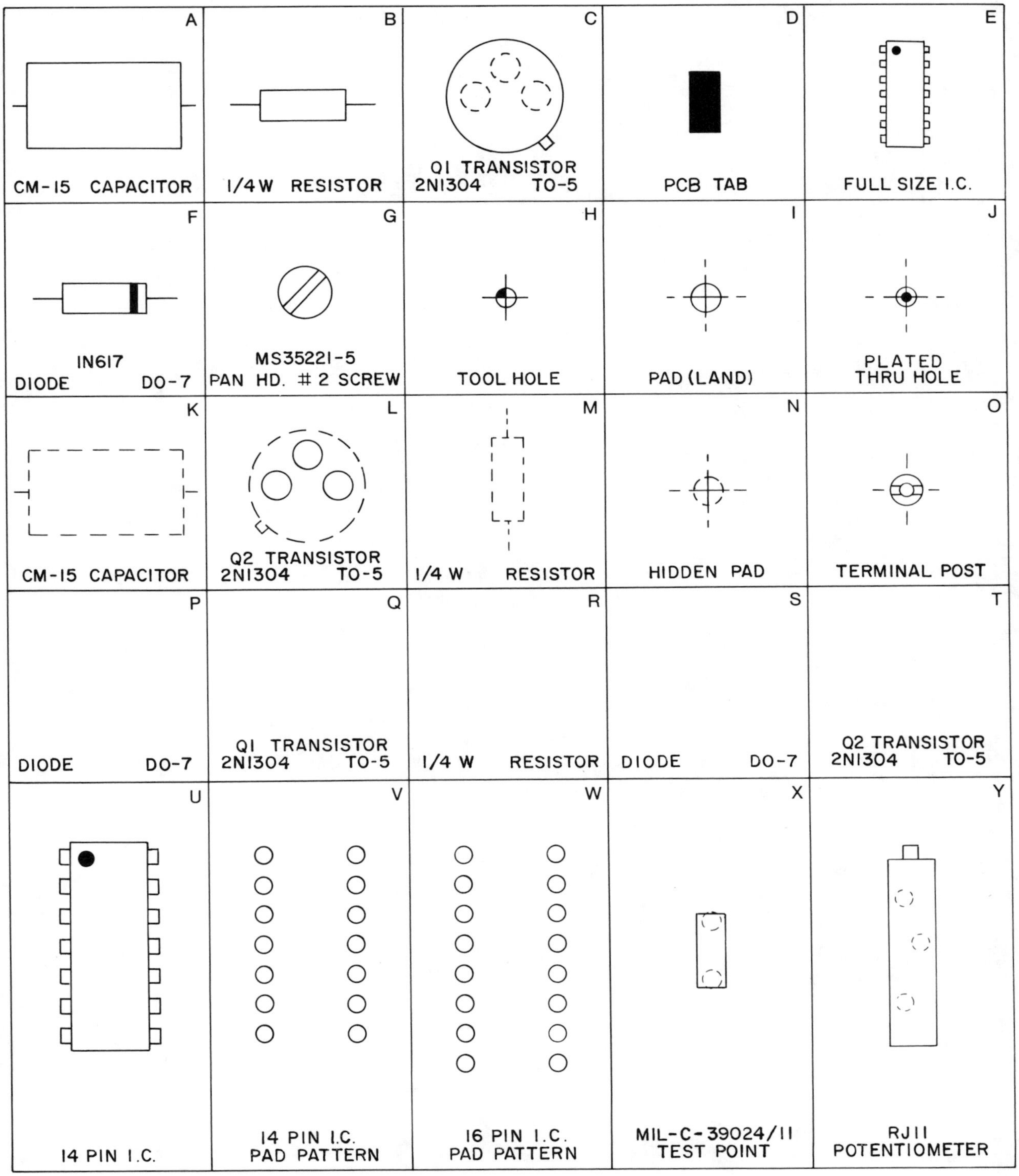

A	B	C	D	E
CM-15 CAPACITOR	1/4 W RESISTOR	Q1 TRANSISTOR 2N1304 TO-5	PCB TAB	FULL SIZE I.C.
F	**G**	**H**	**I**	**J**
1N617 DIODE DO-7	MS35221-5 PAN HD. # 2 SCREW	TOOL HOLE	PAD (LAND)	PLATED THRU HOLE
K	**L**	**M**	**N**	**O**
CM-15 CAPACITOR	Q2 TRANSISTOR 2N1304 TO-5	1/4 W RESISTOR	HIDDEN PAD	TERMINAL POST
P	**Q**	**R**	**S**	**T**
DIODE DO-7	Q1 TRANSISTOR 2N1304 TO-5	1/4 W RESISTOR	DIODE DO-7	Q2 TRANSISTOR 2N1304 TO-5
U	**V**	**W**	**X**	**Y**
14 PIN I.C.	14 PIN I.C. PAD PATTERN	16 PIN I.C. PAD PATTERN	MIL-C-39024/11 TEST POINT	RJ11 POTENTIOMETER

ELECTRONIC COMPONENT MENU #ELCOM1

NOTE 1

Menu entries P, Q, R, S, and T to be made by student as per exercise of CAD-6 (page 89.)

NOTE 2

Unless otherwise specified all components 2 × size.

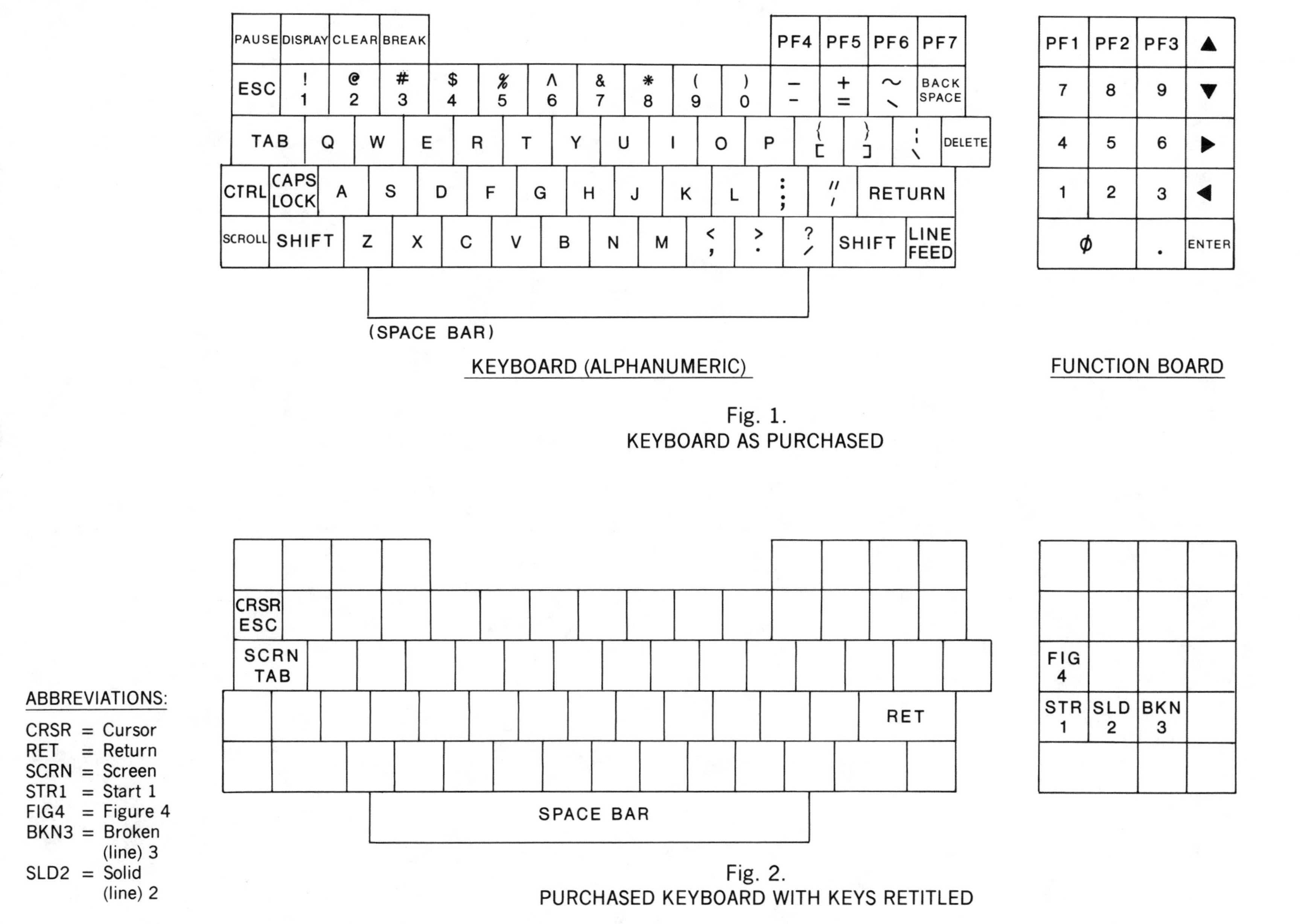

Fig. 1.
KEYBOARD AS PURCHASED

Fig. 2.
PURCHASED KEYBOARD WITH KEYS RETITLED

ABBREVIATIONS:

CRSR = Cursor
RET = Return
SCRN = Screen
STR1 = Start 1
FIG4 = Figure 4
BKN3 = Broken (line) 3
SLD2 = Solid (line) 2

TASK NO.	TASKS AND ROUTINES	EXPLANATION
1-4	START-UP (or LOG-IN) 1. Turn on power	See Fig. 1, CAD-1 — The switch is in the left rear of CRT.
	2. Type in date and press RET.	For example, 21 Dec. 1985. Then press the single key RETURN once (for simple notation, a single key with more than one character, will be underlined). See Note I.
	3. Type in time and press RET.	For example, 8:30 AM. Both steps 2 and 3 are in response to "prompt" requests which appear on the screen.
	4. Press RET twice and wait for prompts to end.	This puts you in "command mode" and cursor (+) blinks on the screen. This ends start-up. See Note II.
5	PAPER SIZE 5. Type @ASIZE and press RET.	This tells plotter that dwg will be A size (8½ × 11). For B size dwg, type @BSIZE.
6	ACTIVATE SCREEN 6. Press TAB (SCRN)	This activates screen; sometimes TAB key is covered with SCREEN title. If step 5 precedes, "A" size dwg outline appears on screen.
7	GRID — LOWER LEFT ORIGIN 7. Type GR and press RET	This displays an X-Y grid on screen's left and bottom edges. See dwg, page 91, Fig. 1. See Note III.
8	ACTIVATE CURSOR 8. Press ESC (Cursor)	This displays the cursor crosshairs on the screen. See Note III.
9	LINE — SINGLE, SOLID (IGM)* 9. Press 1 at 1st pt., JS, and press 2 at 2nd pt.	This creates a solid line as shown in sketch. 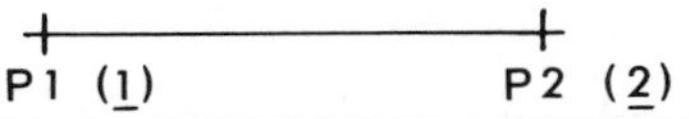 P1 (1) P2 (2)
10	LINE — CONTINUES (IGM)* 10. Press 1 at 1st pt., JS, press 2 at 2nd pt. and JS, PRS 2 at 3rd pt. and JS, PRS 2 at 4th pt.	This creates a solid line. JS (joystick) from pts. 1 to 2 to 3 to 4 as shown in sketch. 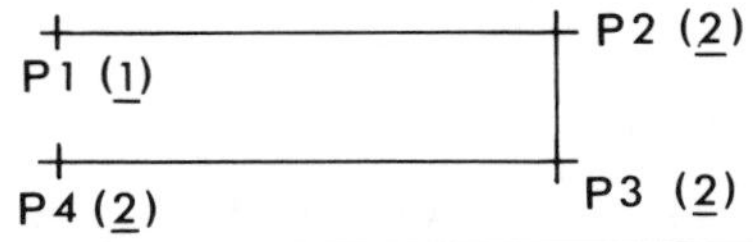 P1 (1) P2 (2) P4 (2) P3 (2)
11	LINE — BROKEN (HIDDEN) (IGM)* 11. Press 1 at P1; JS and press 3 at P2	This creates a broken (dashed) line as shown. See Note IV. 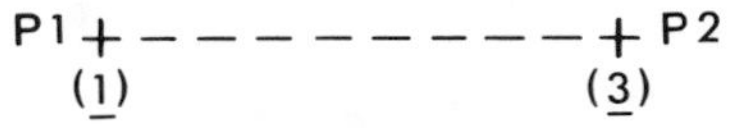 P1 (1) P2 (3)

ABBREVIATIONS, ETC.

CRT = Cathode Ray Tube
RET = Return key
JS = Joystick/move cursor crosshairs by joystick
PRS = Press = Hit one key
PT(S) = Points
P1 & P2 = Point 1 & point 2
(1) & (2) = Press "1" & "2" keys
SOLID LINE = Visible line
BROKEN LINE = Hidden line
*IGM = IN GRAPHICS MODE (See Note III)

Note I: "press" means hit one key—"type" means hit more than one key.

Note II: All subsequent routines to follow will presume that start-up steps 1-4 have been executed properly.

Note III: To switch from "command mode" (e.g., to direct system to execute foregoing grid, paper size, etc.) to "graphics mode" (permits CRT input directions to be displayed), press RET twice or ESC (Cursor = +) once. To return to "command mode", press RET twice (R/R).

Note IV: To go from broken line, e.g., to solid line—press RET once since you are still in graphics mode.

TASK NO.	TASKS AND ROUTINES	EXPLANATION
12	RECTANGLE — SOLID 12. R/R* Press R PRS cursor Press (2) at P1, JS & Press (2) at P2	To create a solid rectangle: Press R/R for command mode; press R for rectangle and press (2) key at diagonal corners P1 and P2 as shown. 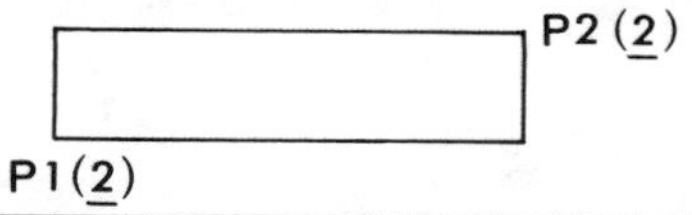 P2 (2) P1 (2)
13	RECTANGLE — BROKEN 13. R/R* Press R PRS cursor Press (3) at P1, JS & Press (3) at P2	To create a broken rectangle: Press R/R for command mode; press R for rectangle and press (3) key at diagonal corners P1 and P2 as shown.  P2 (3) P1 (3)
14	CLEAR SCREEN 14. R/R* Press C Press RET Press Y Press RET Type DP Press RET	To clear the screen, e.g., the rectangle in Fig. 1: Press R/R for "command mode"; Press C key for clear, then press RET; screen prompt will ask, "Do you want to clear the screen?" (Fig. 2). Press Y for yes, and press RET; now type DP for display and press RET—screen clears as in Fig. 3. (See note 5 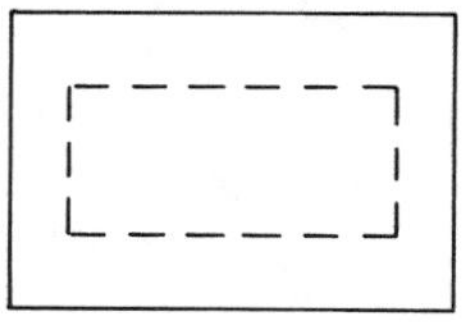) Do you want to clear the screen? Yes. Fig. 1  Fig. 2 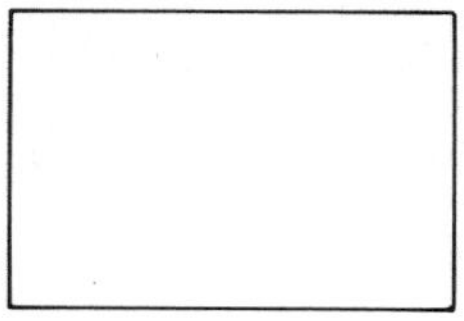 Fig. 3
15	SAVE (STORE) A DWG. 15. Press R/R Type S,PK9005.DWG Press RET	Let's suppose you have just drawn the puck (page 85) on the screen and want to save (or store) it. First press R/R (RET twice) to go from graphics to command mode. Then type S, for save; PK for puck; 9005 for puck #9005; and .DWG for the drawing menu/file (to distinguish it from other menus such as ELSYM1 or ELCOM1). Press RET and puck drawing is now in storage (identified as PK9005.DWG).
16	RECALL (LOAD) A DWG. 16. Press R/R Type L,PK9005.DWG Press RET	To recall (or load) the puck drawing (PK9005.DWG), press RET or R/R key to achieve "command mode" (with blinking cursor). Then type L, for load; PK, 9005, etc.—same I.D. as above. Press RET, and puck drawing is now on screen.
17	RECALL ELECTRONIC SYMBOLS 17. Press R/R Type@ELSYM1 PRS RET Press Cursor JS to P1 Press A Press 4	If you are drawing a logic diagram and wish to recall an amplifier symbol, press R/R to achieve command mode. Then type @ELSYM1 to alert the electronic symbol menu (Appendix I); press RET to end that command; press cursor to see crosshairs on screen (and switch to graphics mode); JS to amplifier "h" point, P1 (5,8)—see Fig. 1 below; press A for amplifier I.D.; and press 4 to recall the figure. The logic symbol will appear on the screen as shown in Fig. 2. 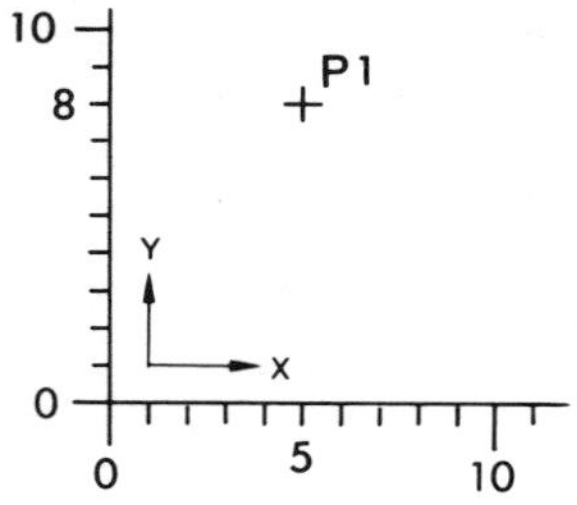 10, 8, 0, 5, 10, P1, Y, X Fig. 1. P1 at X = 5, Y = 8 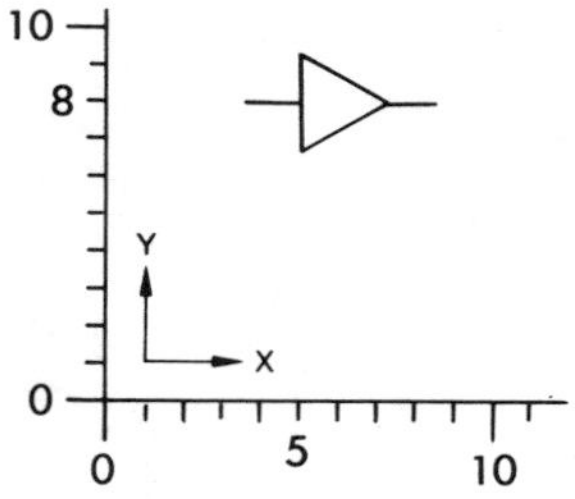 10, 8, 0, 5, 10, Y, X Fig. 2.

ABBREVIATIONS USED

*R/R (begin) = Hit RET twice for command mode.

5 **Note 5:** The RET after each command means "end of command" or period. (see Note III)

CAD TASKS AND ROUTINES APPENDIX M

TASK NO.	TASKS AND ROUTINES	EXPLANATION
18	RECALL COMPONENT SYMBOLS 18. Press R/R Type @ELCOM1 PRS RET Press Cursor JS to P1 Press V Press 4	If you are drawing an I.C. layout and want to recall a 14-pin I.C. pad pattern (V of Appendix J), the routine is similar to Task 17. Press R/R to achieve command mode. Type @ELCOM1 to alert the electronic components menu; press RET to end that command; press cursor to see crosshair on screen (and switch to graphics mode); JS to V pattern "h" point (top left pad), P1 (see Fig. 1); press V for pattern I.D.; and press 4 to recall the figure. The pad pattern will appear on screen as shown in Fig. 2. Fig. 1 P1 at X = 6, Y = 8 Fig. 2.
19	DELETE 19. Press R/R Type DE PRS Cursor JS to P1 press (1) JS to P2 press (1) R/R Type PE PRS RET	To delete a line $\overline{EF}$ of Fig. 1: Press R/R or RET to achieve command mode; type DE for "define edit"; press cursor for crosshair and graphics mode; JS to P1, press (1) on one side and close to $\overline{EF}$; JS to P2, press (1) on other side and close to $\overline{EF}$ to define delete "window". R/R for command mode; PE for "perform edit" and with RET the $\overline{EF}$ line is deleted as in Fig. 2. (This routine is applicable for both single and double lines; hidden or solid.) Fig. 1 Fig. 2
20	ERASE 20. Press R/R Type DE PRS Cursor JS to P1 press (3) JS to P2 press (1) R/R Type PE PRS RET	To erase Q1 of Fig. 1: press R/R or RET to achieve command mode; type DE for "define edit"; press cursor for crosshair and graphics mode; JS to P1, press (3) to define one diagonal end pt. of erasure rectangle; JS to P2, press (1) to define the other diagonal end pt.; R/R for command mode; PE for "perform edit" and with RET the Q1 inside the erasure rectangle disappears as in Fig. 2. Fig. 1 Fig. 2

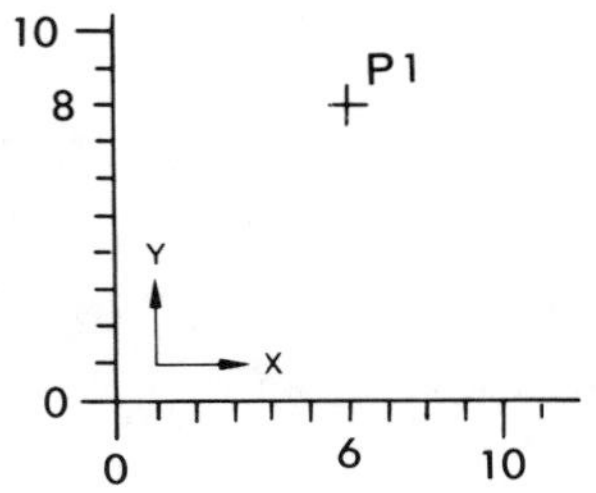

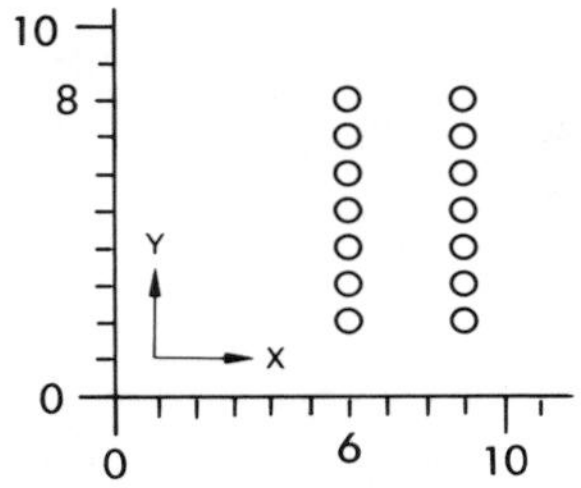

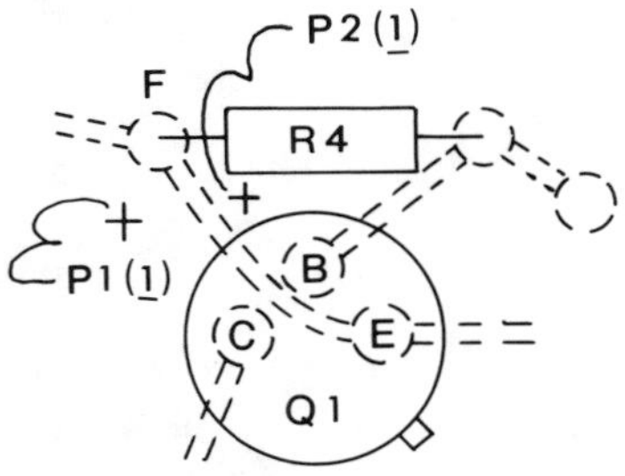

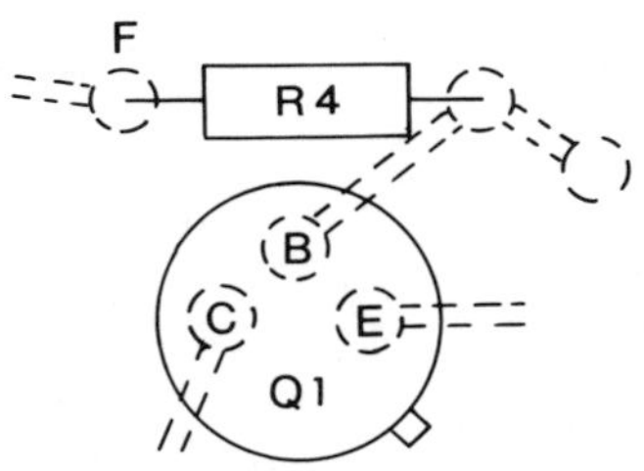

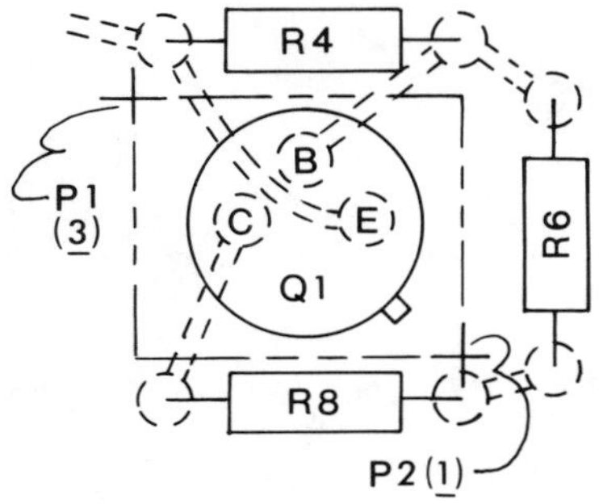

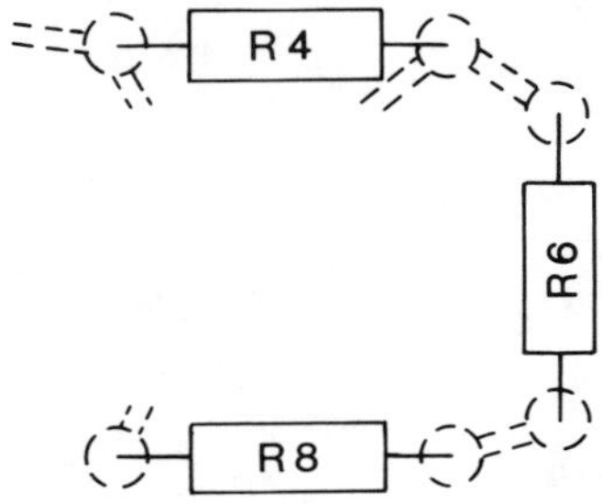

TASK NO.	TASKS AND ROUTINES	EXPLANATION

21 DOUBLE LINE

21. Press R/R
    Type DB,.031,1
    Press Cursor

JS to P1	press (1)
JS to P2	press (2)
JS to P3	press (2)
JS to P4	press (2)
JS to P5	press (1)
JS to P6	press (2)
JS to P7	press (1)
JS to P8	press (3)
JS to P9	press (3)
JS to P10	press (3)
JS to P11	press (1)
JS to P12	press (3)

Press R/R

To draw double lines as shown below: Press R/R for command mode; type DB, for double lines; .031, for center to edge width; 1 for *no* center line; press cursor for crosshair and graphics mode; JS to pts. 1,2,3,4,5, and 6 pressing keys (1),(2),(2),(2),(1), and (2) respectively for *solid* double lines. JS to pts. 7,8,9,10,11, and 12 pressing keys (1),(3),(3),(3),(1), and (3) respectively for *broken* double lines. To complete last step P11 to P12, press R/R.

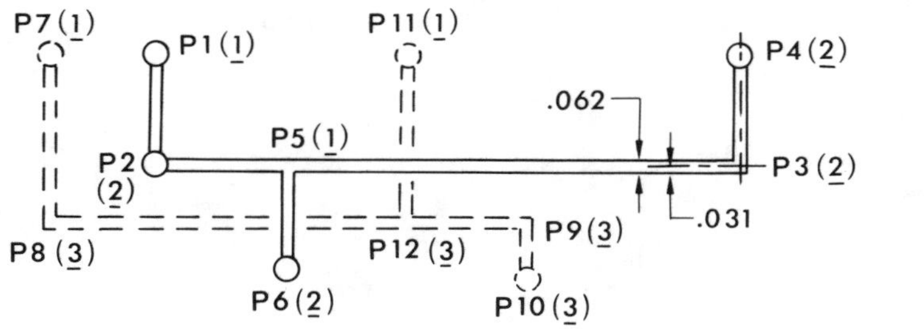

P7 (1) P1 (1) P11 (1) P4 (2) .062 P5 (1) P2 (2) P3 (2) .031 P8 (3) P12 (3) P9 (3) P6 (2) P10 (3)

22 CIRCLE

22. (Fig. 0)
    R/R

Type CC,0,	PRS Cursor
JS to P1	press (2)
JS to P2	press (2)

(Fig. 1)

Type CC,1,	PRS Cursor
JS to P1	press (2)
JS to P2	press (2)

(Fig. 2)

Type CC,2,	PRS Cursor
JS to P1	press (2)
JS to P2	press (2)
JS to P3	press (2)

To draw a .80 dia. circle, Fig. 0: R/R for command mode; type CC for circle; 0 for Fig. 0 style circle; and press cursor for crosshairs and graphics mode. JS to P1 to locate circle center and press (2). JS to P2 to locate .40 radius and press (2). Fig. 0 circle now appears. Figures 1 and 2 are made as indicated.

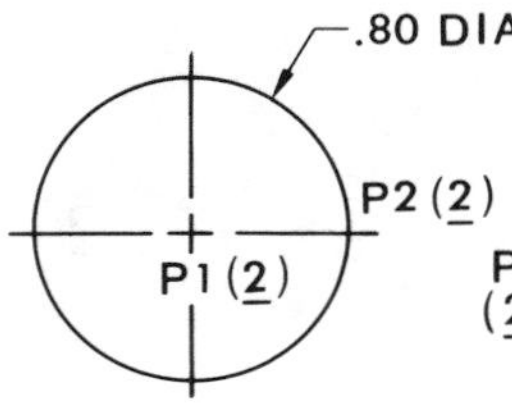

.80 DIA P2 (2) P1 (2)

Fig. 0

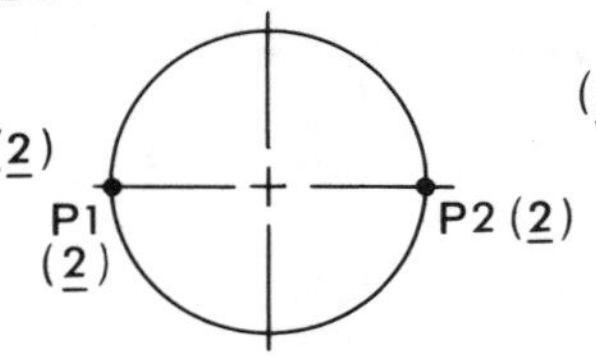

P1 (2) P2 (2)

Fig. 1

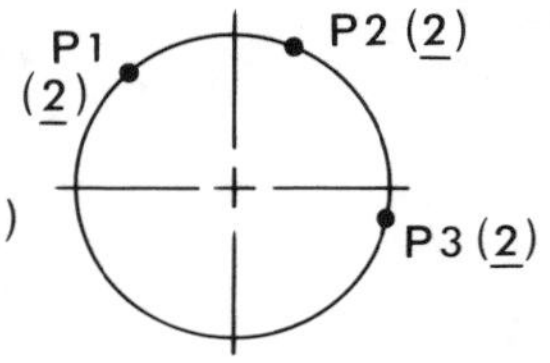

P1 (2) P2 (2) P3 (2)

Fig. 2

Note: Circle styles of Figs. 1 and 2 require *no* prior known center or radius.

23 ARC

23. (Fig. 10)
    R/R

Type CC,10,	PRS Cursor
JS to P1	press (2)
JS to P2	press (2)
JS to P3	press (2)

(Fig. 11)

Type CC,11,	PRS Cursor
JS to P1	press (2)
JS to P2	press (2)

(Fig. 12)

Type CC,12,	PRS Cursor
JS to P1	press (2)
JS to P2	press (2)
JS to P3	press (2)

To draw the arc of Fig. 10: R/R for command mode; Type CC for circle; 10 for Fig. 10 style arc; and press cursor for crosshair and graphics mode. JS to P1 to locate center and press (2); JS to P2 to locate one end of arc and press (2); JS to P3 at the end of arc, and press (2). Figure 10 now appears.
Figures 11 and 12 are made as indicated.

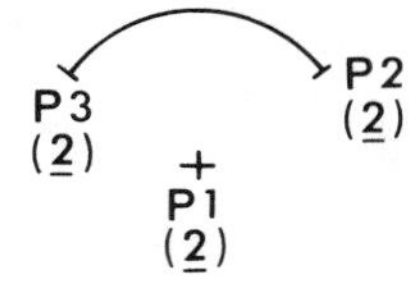

P3 (2) P2 (2) P1 (2)

Fig. 10

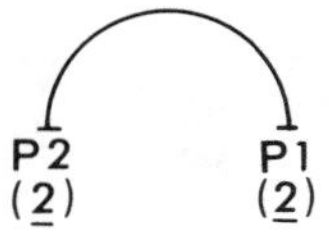

P2 (2) P1 (2)

Fig. 11

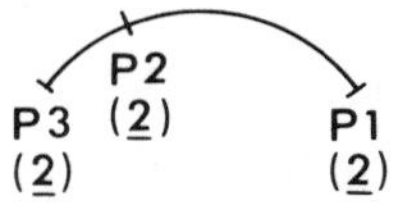

P2 (2) P3 (2) P1 (2)

Fig. 12

Note: Arc styles of Figs. 11 and 12 require *no* prior known center or radius.

TASK NO.	TASKS AND ROUTINES	EXPLANATION
24	CROSSHATCH 24. Press R/R Type CH,LINES.PTN,.06,45 Press Cursor JS to P1 Press (1) JS to P2 Press (2) JS to P3 Press (2) JS to P4 Press (2) JS to P5 Press (1) JS to P6 Press (2) JS to P7 Press (2) JS to P8 Press (2) Press R/R — — — — — — — — — — R/R (for counter-crosshatch) Type CH,LINES.PTN,.06,135 Press Cursor JS to P9 Press (1) JS to P10 Press (2) JS to P11 Press (2) JS to P12 Press (2) Press R/R — — — — — — — — — — CROSSHATCH FILL-IN R/R (for fill-in; see Fig. 2) Type CH,NET3.PTN.01 Press Cursor JS to P13 Press (1) JS to P14 Press (2) JS to P15 Press (2) Press R/R	To crosshatch the processor housing of Fig. 1: R/R for command mode; type CH, for crosshatch; LINES. PTN, for a lines pattern; .06 for lines .06 apart; 45 for lines 45° counterclockwise from X axis; press cursor for crosshairs and graphics mode. JS to P1 and press (1) to begin to define outermost boundary; JS to P2 and press (2) to define 2nd pt. in clockwise sequence; continue with P3 and P4 clockwise to complete outer boundary and press (2) at each pt. The inner boundary, P5, P6, P7, and P8 is now identified by (1),(2),(2), and (2) respectively. Finally the counter-crosshatch area P9, P10, P11, and P12 is given a similar command: CH,LINES.PTN,.06,135 except that the angle is 135° instead of 45°. After P8 (2) and R/R, crosshatching begins; after P9 (1) to P12 (2), press R/R and counter-crosshatching begins. P1 (1) P2 (2) .06 45° P5 (1) P9 (1) P10 (2) P6 (2) P8 (2) P12 (2) P11 (2) P7 (2) P4 (2) P3 (2) 135° Fig. 1 To crosshatch-**fill-in:** Use same routine as above with very close linework (use ".01" inches apart). For faster fill-in use a dense line pattern like NET3.PTN (see Fig. 3). P13 P14 P15 .01 .01 Fig. 2 Fig. 3
25	ROTATE 25. First draw, save, and/or recall figure, then: R/R Type RT,F,90 Press RET PRS Cursor JS to P1 Press P Press 4 **Note:** After all rotating operations are completed, NORMALIZE succeeding commands by: R/R Type RT, F, N Press RET	To rotate Fig. 1, 90° counterclockwise (CCW): First draw, save, and/or recall Fig. 1 (Task no. 17, with letter P for capacitor—see Appendix I). Then R/R for command mode; Type RT, for rotate; F, for figure; 90 for 90° CCW; press RET; Press CURSOR and JS to P1; Press P for capacitor and press 4 to recall the figure. Figure 1 is now rotated into Fig. 2. **Note:** Point "h" is the fulcrum for rotation. After all rotating operations are completed it is *important* to *normalize.* See *normalize* command on left. P1 h P1 h Fig. 1 Fig. 2

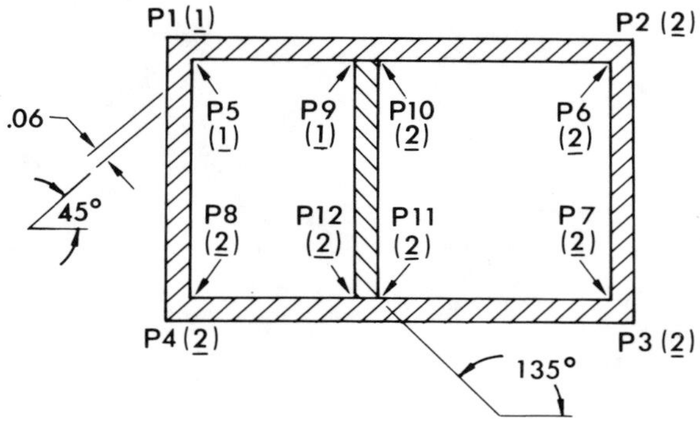

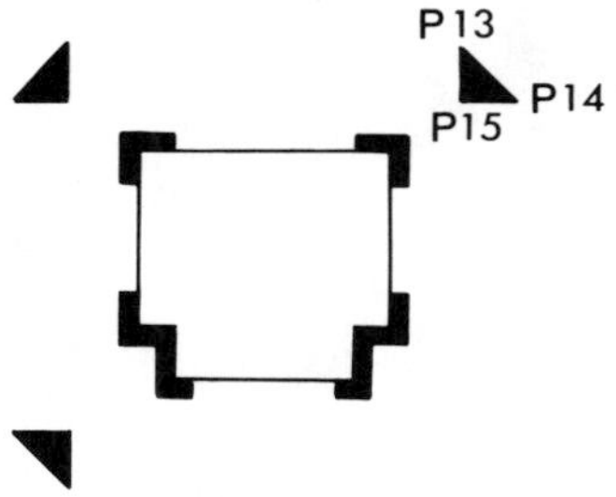

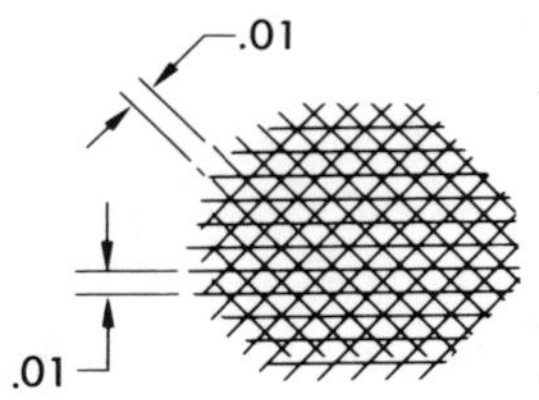

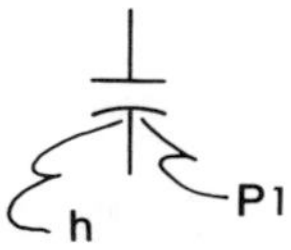

TASK NO.	TASKS AND ROUTINES	EXPLANATION
26	STEP AND REPEAT 26. (Draw or recall dwg., e.g., by "L,PCB001.DWG") RET Type @ELCOM1 RET Press D RET Type SR,S,N,.3 Press Cursor JS to P1 press (1) JS to P2+ press (1) (Tabs now appear as in Fig. 1.)	To place PCB **tabs** repeatedly from electronic components menu #ELCOM1 item "D" by the step and repeat process after PCB001 .DWG is either drawn or recalled: RET for end of command; Type @ELCOM1 to activate ELCOM1 menu; press "D" for item D (PCB tab); type SR, for step and repeat; S for "spanned" (between P1 and P2+); N for normal direction; .3 for .3″ apart (between "h" pts.); press cursor to activate crosshair and graphics mode; JS to P1 (the "h" of left-most tab) and press (1); JS to P2+ (just slightly to the right of P2, beyond the "h" of the right-most tab) and press (1). The tabs now appear as in Fig. 1. P2+ (1) P1(h) (1) P2(h) .3 (TYP) Fig. 1 (2 × size) (#PCB001.DWG plus tabs)
27	MIRRORING 27. R/R Type @PAGE2 RET Press Cursor Now draw Figure 2. Press R/R Type S,MIRRO1.SCH Press RET Type F,MIRRO1.SCH,B Press RET Type RT,F,N RET Type RT,F,Y180 RET Press Cursor JS to grid origin (0,0) PRS B PRS 4 (Figure 3 now appears on the screen with all lines shown solid. The grid is still there but will not reappear when Fig. 3 is recalled from storage or plotted.)	To "mirror" Fig. 2 into Fig. 3: Press R/R for command mode; type @PAGE2 and RET to obtain Fig. 1 grid in screen center; press cursor for graphics mode and to draw Fig. 2. Now CAD-DRAW Fig. 2. Press R/R for command mode; type S,MIRRO1.SCH to save Fig. 2, now identified as MIRR01.SCH; press RET to end command and store Fig. 2. Type F,MIRR01.SCH,B and press RET to store Fig. 2 in the "B" box of the SCH menu as a figure which can be modified ("B" box is used if "A" box is occupied and needed). Type RT,F,N and RET to establish that Rotation is about to take place on the Figure (2) now stored in its "normal" orientation. Type RT,F,Y180 and press RET to rotate FIG about the Y axis, 180°. Press cursor to JS to grid origin (0,0), where Y axis acts as a ℄ — fulcrum for mirroring. Press B and 4 to activate the "B" box of SCH menu and bring it to the screen rotated 180°. The screen now shows Fig. 3 with all lines solid.
27b	MIRROR STORAGE 27b. (When Fig. 3 appears on screen) R/R Type S,MIRRO2. SCH Press RET (Figure 3 is now stored as MIRRO2.SCH)	To store Fig. 3: R/R for command mode. Type S,MIRR02.SCH to save Fig. 3 in the SCH file. At press RET, the Fig. 3 is put in the SCH file. (0,0) ℄ Fig. 1 Fig. 2 Fig. 3

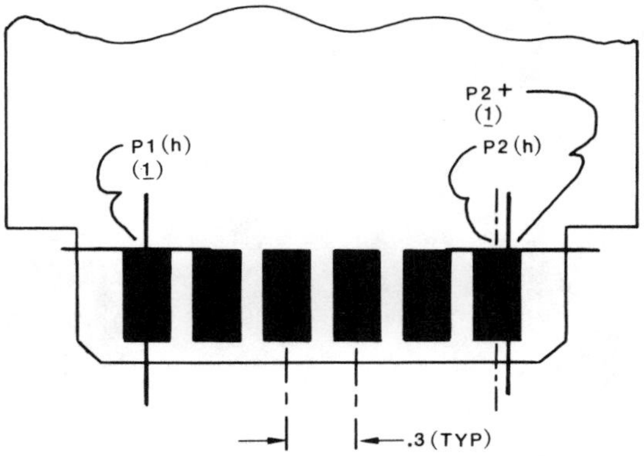

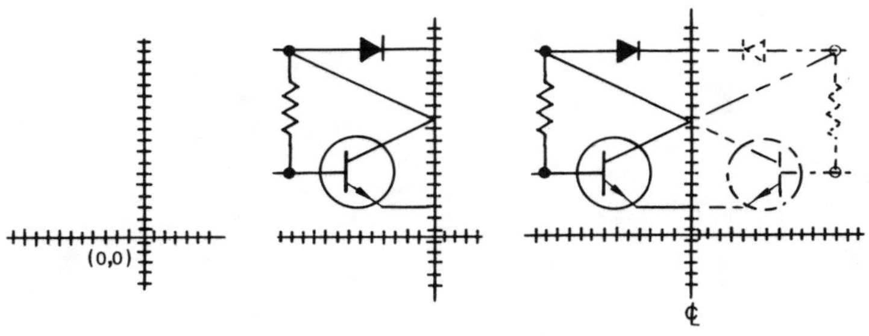

TASK NO.	TASKS AND ROUTINES	EXPLANATION

28

DIMENSIONING

28. To CAD-dim. the outline dwg of Fig. 1: First CAD-draw the outline, or recall it; then to execute DIM A:

R/R

Type DM,D,	Press cursor
JS to 1a	Press (1)
JS to 2a	Press (2)
JS to 3a	Press (2)
JS to 4a	Press (2)

Note: 2a is any pt. along line 1a-4a; 3a is any pt. along dim. line A. Dim. A appears after: 4a press (2),

To execute DIM. B:

JS to 1b	Press (1)
JS to 2b	Press (2)
JS to 3b	Press (2)
JS to 4b	Press (2)

Note: 2b is any pt. along line 1b-4b; 3b is any pt. along dim. line B. Dim. B appears after: 4b press (2).

Dims. C & D executed like dims A & B.

To execute DIM. E: (see Fig. 2)

JS to 1e	Press (1)
JS to 2e	Press (2)
JS to 3e	Press (2)
JS to 4e	Press (2)

Note: 3e can be on either side of close dim. E arrows, wherever dim. E is pref. dim. E appears after: 4e press (2).

DIM. F executed like dim. E.

Figure 3 is a chart of full size, two decimal CAD dim. — arrow limits; to suggest when C/D type dims will occur or when a 3e/3f type JS movement is required.

EXAMPLE FOR HOLE DIMS

R/R for command mode; Type DM for dim; D for decimals; cursor for crosshair and graphics mode; 1a(1) for first end pt.; 2a(2) for dim direction; 3a(2) for length of first ext. line and dim. placement; 4a(2) for second end pt. Dim A now appears.

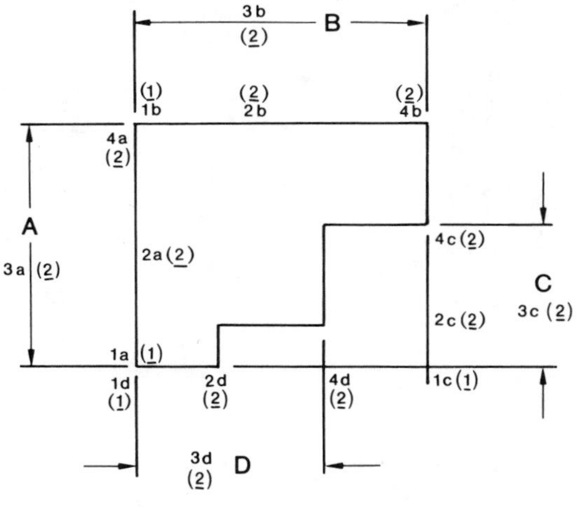

Fig. 1

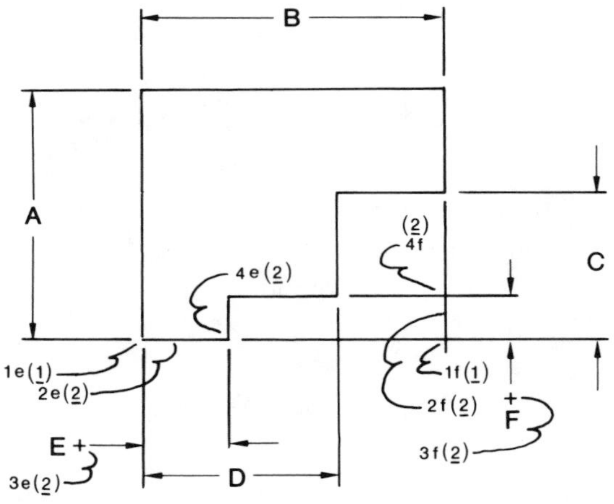

Fig. 2

Fig. 3

FULL SIZE — 2 DECIMAL CAD DIM. — ARROW LIMITS

VERTICAL DIMS.	A	.60 + = DIM & ARROWS INSIDE	.xx
	C	.15 – .55 = DIM. IN, ARROWS OUT	.xx
	F	.10 OR LESS = DIM. & ARROWS OUT	.xx

HORIZ. DIMS.

B	D	E
.70 +	.40 — .65	.35 or less
.xx	.xx	.xx

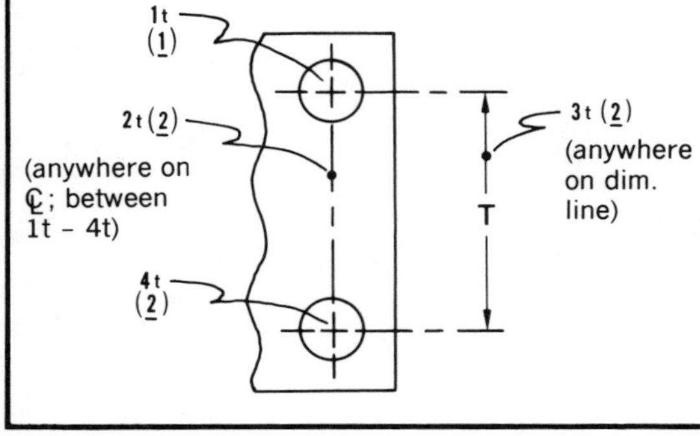

APPENDIX R

CAD TASKS AND ROUTINES

TASK NO.	TASKS AND ROUTINES	EXPLANATION
29	SCALE (For full scale Fig. 1—down .5; recall Fig. 1 by "L,PCB001.DWG") RET 29. Type SC,F,.5 RET (Fig. 2 without dims appears) — — — — — — — — — — Fig. 2 can now be dimensioned at full size. For 2 × size Fig. 2 with dimensions (PCB002.DWG) 29. R/R Type SC,F,2 RET (Fig. 2 with dims now appears—but 2 × size) — — — — — — — — — — To dim Fig. 1 with true size dims as in Fig. 3; After recall of Fig. 1 (Type L,PCB001.DWG RET) 29. Type SC,F,2 RET Type DM,D PRS cursor JS to P1 PRS (1) JS to P2 PRS (2) JS to P3 PRS (2) etc., see DIM. Task 28	To reduce Fig. 1 from 2 × size to full size: Recall file # (PCB001.DWG) to screen with a RET; Type SC,F,.5; SC, for scale; F, for figure; .5 for .50 reduction; RET for end of command. Fig. 2 w/o dims now appears. If Fig. 2 is now to be revised &/or dim., the new dwg can be doubled again with: Type SC,F,2 RET. To dim a 2 × scale dwg with full size (true) dims: Type SC,F,2 RET. And type DM,D Prs cursor. Fig. 1 — 2 × Size (#PCB001.DWG) Fig. 2 — Full Scale (#PCB002.DWG) Fig. 3 (#PCB003.DWG)
30	TEXT (lettering) After drawing outline is recalled to screen (#PK9002.DWG) 30. R/R Type T,CTF,STDFNT,.12, .5,2,0,1 Press cursor JS to P1 PRS (1) JS to P2 PRS (1) Type .20 × .20 (5.08 × 5.08) CHAMFER 4 PLCS R/R Type T,CBF,STDFNT,.12, .5,2,0,1 Press cursor JS to P3 PRS (1) JS to P4 PRS (1) Type BUTTON .50 SQ (12.7 SQ) 4 PLCS R/R	Assuming that the puck below has already been CAD-dimensioned in both decimal inches and millimeters, where capable on a preceding layer, the below remaining text callouts could be entered on a separate layer after the puck (dwg #PK9002.DWG-without dim.) is recalled to the screen. Press R/R for command mode; Type T for text; CTF for center, top fitted; STDFNT for standard font (standard lettering); .12 for letter ht.; .5 for .5 of ht. horiz. spacing bet. letters; 2 for approx. dbl. space vert. line spacing; 0 for Zero deg. slant; and 1 for aspect ratio (letter width ÷ letter ht. = 1). JS to P1 and P2 to locate top line and length for top line to be centered within and/or fitted. The remaining callouts have similar commands and most CAD systems are programmed with "defaults" so that the operator only types the first two command "modifiers" (e.g., "T,CBF" or just "T"). However the system can automatically place or "justify" all lettering starting at left top (LT) or end all lettering at rt. top (RT) as at P6 and P5 respectively. **Note:** CBF = center bottom fitted CMF = center middle fitted **NOTE:** Ref. to P6 .80 DIA in inches (20.32 DIA) in MM FRONT VIEW

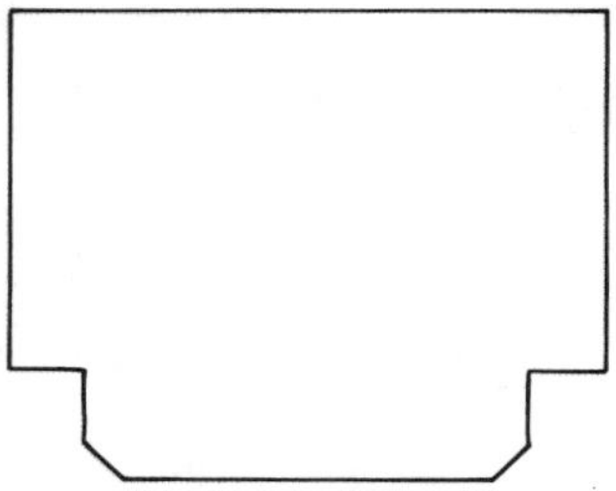

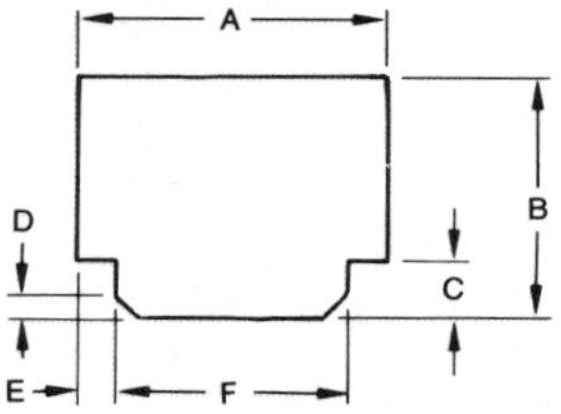

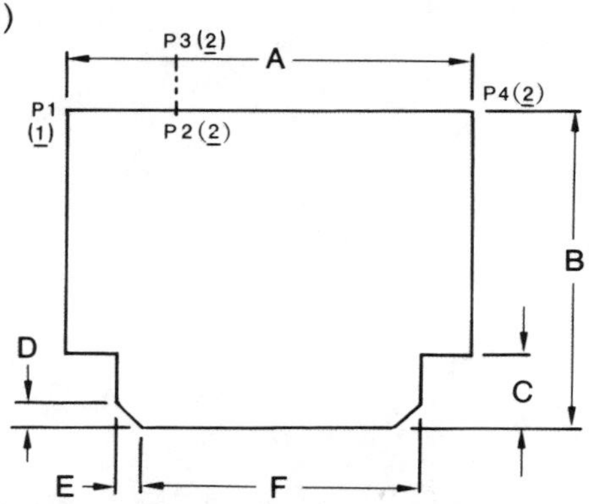

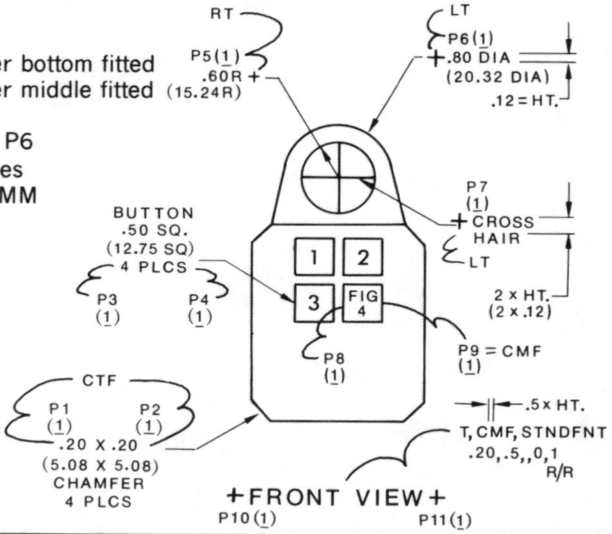

TASK NO.	TASKS AND ROUTINES	EXPLANATION
31	ZOOM-IN (DISPLAY-ENLARGED) 31. R/R Type DP PRS cursor JS to P1 PRS (1) JS to P2 PRS (1) (Fig. 2, w/o P3–P4 erasure rectangle, now fills screen)	To zoom (enlarge) a portion of a drawing as in Fig. 1: R/R for command mode; type DP for display press cursor for graphics mode and crosshairs; JS to P1 and press (1) to define one end pt. of the zoom rectangle; JS to P2 and press (1) to define the other end of the zoom rectangle. The zoomed area now appears as in Fig. 2—occupies whole screen. Zoomed Fig. 2 can now be modified, e.g., with P3–P4 erasure and P5–P6 delete. For erasure, see task 20. For delete see task 19. In each case above the PE (perform edit) function allows the viewer to see the modifications while the Fig. 2 is still enlarged (see Fig. 3). Additionally, the Fig. 3 can be "zoomed out" (reverse of zoomed in) to become original size of Fig. 1, as in Fig. 4 with recalled pad and added conductors. For recall of components, see task 18. Fig. 1 Fig. 2 Fig. 3 Fig. 4
31a	ZOOM OUT (Enlarged display returned to smaller original—from Fig. 3 to size of Fig. 4) 31a. R/R Type DP PRS RET	
32	PLOTTING 32. R/R PL Type PL, PK9005.DWG Press RET	Let's assume that the dimensioned puck of page 85 has been completed and stored as: #PK9005.DWG. To plot this drawing using full scale and the current working pen: R/R for command mode; type PL, for Plot; PK9005.DWG for puck; and RET for end of command. The plotter now begins plotting the puck.